GUIDE TO
POST-EARTHQUAKE INVESTIGATION OF LIFELINES

by the
Earthquake Investigations Committee
of the
Technical Council on Lifeline Earthquake Engineering
of the
American Society of Civil Engineers

**Technical Council on
Lifeline Earthquake Engineering
Monograph No. 3
August, 1991**

Edited by Anshel J. Schiff

Published by the
American Society of Civil Engineers
345 East 47th Street
New York, New York 10017-2398

ABSTRACT

This committee report, *Guide to Post-Earthquake Investigation of Lifelines,* will help the investigator become familiar with the overall operation of major lifeline systems, with the function and operation of lifeline facilities and equipment, with past seismic performance, and with methods to gather pertinent information. The beginning chapters describe how to prepare for a post-earthquake investigation and provide a summary of phenomena related to earthquakes and their effect on lifelines. Chapters 6-11 explain typical system configurations and overall operation of the following lifelines: power, water, sewage, transportation, communications, liquid fuel and natural gas sytems. System facilities and equipment are described for each lifeline, including their role in overall system operation and their seismic performance. Detailed guidance is provided for their investigation. Chapter 12 discusses tanks and emergency power, facilities common to many lifelines. The appendices present check lists in a form that can be used as field guides during investigations. They also suggest formats for reconnaissance reports, tips on technical report writing, and references to reconnaissance reports.

Library of Congress Cataloging-in-Publication Data

American Society of Civil Engineers. Earthquake Investigations Committee.
    Guide to post earthquake investigations of lifelines/by the Earthquake Investigations Committee of the Technical Council on Lifeline Earthquake Engineering of the American Society of Civil Engineers; edited by Anshel J. Schiff.
        p.   cm.—(Technical Council on Lifeline Earthquake Engineering monograph: no. 3)
    Includes bibliographical references.
    ISBN 0-87262-844-2
        1. Earthquake engineering. 2. Underground utility lines—Earthquake effects. 3. Public utilities. 4. Transportation. I. Schiff, Anshel J. II. Title. III. Series.
TA654.6.A434   1991
624.1'762—dc20   91-24240
                                                                               CIP

      The material presented in this publication has been prepared in accordance with generally recognized engineering principles and practices, and is for general information only. This information should not be used without first securing competent advice with respect to its suitability for any general or specific application.
      The contents of this publication are not intended to be and should not be construed to be a standard of the American Society of Civil Engineers (ASCE) and are not intended for use as a reference in purchase specifications, contracts, regulations, statutes, or any other legal document.
      No reference made in this publication to any specific method, product, process, or service constitutes or implies an endorsement, recommendation, or warranty thereof by ASCE.
      ASCE makes no representation or warranty of any kind, whether express or implied, concerning the accuracy, completeness, suitability or utility of any information, apparatus, product, or process discussed in this publication, and assumes no liability therefor.
      Anyone utilizing this information assumes all liability arising from such use, including but not limited to infringement of any patent or patents.

Authorization to photocopy material for internal or personal use under circumstances not falling within the fair use provisions of the Copyright Act is granted by ASCE to libraries and other users registered with the Copyright Clearance Center (CCC) Transactional Reporting Service, provided that the base fee of $1.00 per article plus $.15 per page is paid directly to CCC, 27 Congress Street, Salem, MA 01970. The identification for ASCE Books is 0-87262/91. $1 + .15. Requests for special permission or bulk copying should be addressed to Reprints/Permissions Department.

Copyright © 1991 by the American Society of Civil Engineers,
All Rights Reserved.
Library of Congress Catalog Card No: 91-24240
ISBN 0-87262-844-2
Manufactured in the United States of America.

FOREWORD

For a number of years, the post-earthquake investigation of lifelines was primarily directed at documenting the location and types of lifeline damage that occurred. As a result, the types of facilities and equipment that were vulnerable to earthquake damage were identified and certain patterns of damage started to emerge. While this provided a start to defining lifeline vulnerabilities, these investigations had several deficiencies. The formally organized investigations, particularly those in foreign countries, typically had one individual who was responsible for collecting all lifeline data. Considering the number and complexity of lifelines and the type of data that was needed, this was an impossible task for one individual. In addition to the type of facilities and equipment that were damaged and a count of each type, information was needed to advance the state-of-the-art of seismic design of facilities and equipment. Failure modes and factors contributing to them had to be identified. Also, the impacts of failures on system operations and the resources and time required to restore facilities was needed for utility emergency planning as well as for emergency response planning within the community. The expertise required for a post-earthquake investigator, even if only one lifeline was to be focused upon, is beyond what any individual within that lifeline would be expected to know. Knowledge of the entire system and its operation is needed as well as the operation, construction and design of system facilities and equipment. Finally, information on the response of the overall system, facilities and equipment to past earthquake should be known.

This training guide is an attempt to gather material to meet four primary objectives.

1. Provide information so that an investigator will be familiar with the overall operation of major lifeline systems.

2. Provide information so that the investigator will be familiar with the function and operation of lifeline facilities and equipment.

3. Provide information so that the investigator will be familiar with past seismic performance of the lifeline and its facilities and equipment.

4. Provide guidance for methods to gather pertinent information and a means for improving the process.

Anshel J. Schiff
May 1991

TABLE OF CONTENTS

	Foreword	iii
	Table of Contents	v
1.	Introduction	1
2.	General Pre-Departure Preparations	15
3.	Post-Earthquake/Pre-Departure Preparations	25
4.	Introduction to Earthquakes and Their Effects	31
5.	General Procedures for Conducting an Investigation	43
6.	Electric Power Systems	53
7.	Water Systems	105
8.	Sewage Systems	115
9.	Transportation Systems	133
	Airports	135
	Ports	141
	Highways	173
10.	Communications Systems	181
11.	Liquefied Fuels and Natural Gas Systems	199
12.	Common Facilities	211
	A. Tanks	213
	B. Emergency Power	219
13.	Acknowledgements	229
14.	Appendices	231
	A. Field Guide for Lifeline Earthquake Investigations	233
	B. Report Format	261
	C. Tips on Technical Writing	263
	D. References to Reconnaissance Reports	267

1. Introduction

   Organization of Guide
   Objective of Post-Earthquake Investigations
   Use of Guide
   Risks Associated with Post-Earthquake Investigations
   Acknowledgement of Risks with an Earthquake
      Investigation
   Lifeline Earthquake Investigations: Past and Future
   The Post-Earthquake Investigation Process

1. INTRODUCTION

ORGANIZATION OF GUIDE

The guide is the start of a living document for educating post-earthquake investigators. The content of each section is described below.

1. Introduction

    The Introduction will state the objectives of post-earthquake investigations, point out the risks associated with investigations, explain how to use the notes, give a brief history of earthquake investigations in the United States, and review the investigation process.

2. General Preparations

    The General Preparations Section describes the things that an investigator should have done prior to being called to investigate an earthquake so that he or she is prepared to make final preparation before leaving for the investigation.

3. Post-Earthquake/Pre-Departure Preparations

    The Post-Earthquake/Pre-Departure Preparations Section describes the tasks that should be done after an earthquake has occurred but before the reconnaissance team has departed. Tasks for the team leader and the individual investigator are covered.

4. General Introduction to Earthquakes and Their Effects

    An introduction to the causes of earthquakes and their effects including faulting, liquefaction, landslides, and subsidence. Earthquake terminology will also be introduced.

5. General Procedures

    The General Procedures Section describes those things that should be done when conducting an investigation.

6. - 12. Lifeline Systems

    The following lifelines are covered.
    Power Systems
    Water Systems
    Sewage Systems
    Transportation Systems
    Communications Systems
    Liquefied Fuels and Natural Gas Systems
    Common Facilities
        Tanks
        Emergency Power Systems

For each lifeline the following topics will be discussed.

> System Configuration and Operation
> System Facilities and Their Functions
> Equipment and Their Functions
> Sources of Damage Information
> Seismic Performance of System Facilities and Equipment
> Guide for Investigating Specific Facilities and Equipment
> Case Study of an Earthquake Investigation

13. Acknowledgements
14. Appendices

> Appendix A. Field Guide for Lifeline Earthquake Investigations
>
> This Appendix contains check lists that are meant to serve as reminders of the way things are to be done and information that is to be gathered. It is to designed to be copied from the notes and to serve as a field guide.
>
> Appendix B. Report Format
>
> This section contains a format suggested for reconnaissance reports.
>
> Appendix C. Tip on Technical Writing
>
> This section provides examples of proper usage of several commonly misused punctuation.
>
> Appendix D. References to Reconnaissance Reports
>
> A list of reconnaissance reports that contain significant sections on lifelines is provided.

OBJECTIVES OF POST-EARTHQUAKE INVESTIGATIONS

The overall objective of post-earthquake investigations is to improve the seismic resistance of lifelines. It is anticipated that this will be accomplished in the following ways:

1. Identify the facilities and equipment that are vulnerable, their failure modes and factors that contributed to the failure. This will allow utility personnel, facilities designers, and equipment manufacturers to develop facilities and equipment which have better seismic performance.

2. Identify the consequences of specific failures in terms of the extent and duration of disruption. This will enable utilities to determine the relative importance of different facilities and components so that equipment and facility designs and system operations can be improved. Also, priorities for mitigation measures can be established.

3. Identify the impacts of lifeline disruption on the emergency response community so that the emergency response community will be better able to prepare for earthquakes.

4. Disseminate observations about lifeline damage, its impact on the extent and duration of lifeline disruption, and its impact on the emergency response community. It is important that utilities be made aware that their facilities are vulnerable to earthquake damage and that the damage can be costly and disruptive. It is important that the technical community be made aware that earthquakes should be given consideration in the design, fabrication, construction, installation, and operation of lifelines.

5. Make recommendations for improving post-earthquake investigation of lifelines and the training of investigators.

To summarize, these objectives provide a focus for investigators: find failures, identify their cause and the factors that contribute to them, identify impact of the failures, and communicate your findings to the appropriate user communities.

USE OF GUIDE

It is suggested that after reading the Introduction, the damage reports from previous earthquakes be reviewed. A list of damage reports is given in Appendix B. This should familiarize the investigator with the types of damage that have been observed and how such damage has been reported. By reviewing the reports in chronological order, the evolution of the character of the reports should also be obvious. It is then suggested that the Training Guide be read, starting with Section 2, General Pre-Departure Preparations. Rereading the damage reports again after the Guide has been read, you will better appreciate the importance of information that has been requested in the Guide. The breadth of material covered in the Guide, even if only one lifeline is considered is very broad. To be fully absorbed, so that its advice can be properly used in the field, the Guide will have to be reviewed several times. In general, it will be impractical to take the Guide in the field as a reference document.

Appendix A, Field Guide for Lifeline Earthquake Investigations, contains several check lists that are meant to be copied and taken and used in the field to serve as reminders of how things are to be done and what data are to be collected.

RISKS ASSOCIATED WITH POST-EARTHQUAKE INVESTIGATIONS

An investigator may be subjected to several hazards when investigating an earthquake.

The investigator may be subjected to additional earthquakes (aftershocks) during the course of the investigation. For this reason, it is unwise to enter severely damaged structures since an aftershock may cause collapse. Also, in selecting housing the seismic resistance of the structure should be given some consideration.

In the aftermath of a major earthquake there are typically demolition and reconstruction activities using heavy equipment to clear rubble and to restore facilities. At any site involving large equipment, care must be exercised to keep out of the way. In the often chaotic situation after an earthquake these risks will be heightened.

Health hazards may be more severe after an earthquake. Water distribution systems may be damaged so that water may not be readily available. Often water and sewage pipes are damaged so that the normally potable water may become unsafe. Also, water treatment facilities may be damaged so that an otherwise functioning water system is not distributing safe water. Sewage piping or treatment facilities may be damaged and inoperable so that raw sewage may be gathering throughout the city creating an additional health hazard. These conditions of poor sanitation carry the additional risk of exposure to communicable diseases.

Should you get injured, the health delivery system may be overwhelmed with other earthquake victims and treatment facilities may themselves be damaged. Thus, it is particularly important to avoid being injured. Finally, the stability of the local political situation may also impact your security.

It must be emphasized that entering an earthquake disaster area just after an earthquake involves many risks. While training notes provide some guidance, they may be incomplete and must be applied with judgement to each particular situation when in the field. Your health and safety is your responsibility. The purpose of the investigation is to gather earthquake statistics, not to become one.

In each earthquake a unique set of circumstances will be encountered so that some of the advice given in this guide may be inappropriate to your situation. While the investigation team will have a leader, his or her responsibility is technical coordination. It is you who must look out for and be responsible for your own safety.

To emphasize your role in providing for your own safety and to make sure that you realize that you understand the hazards associated with an earthquake investigation you will be required to sign the enclosed statement. The signed copy is to be returned to the team leader prior to departing for the investigation.

Acknowledgement of Risks Associated With an Earthquake Investigation

I have read the above description which is indicative and not an exhaustive list of the hazards associated with an earthquake investigation. I understand that my safety and health are my responsibility and not that of the investigation team leader, the Earthquake Investigation Committee, ASCE, EERI, NRC, NAS, or other organization supporting the investigation.

_____   _____
Name   Date

ASCE WAIVER AND RELEASE

Inconsideration of your sponsorship and of your assistance in arranging my trip to _____ , the undersigned, intending to be legally bound hereby for myself, my heirs, executors and administrator, waives and releases any and all rights and claims that I now or may in the future have against the American Society of Civil Engineers ("ASCE"), its directors, officers, employees, agents and representatives, for any and all injuries (including death), damages and expenses suffer of incurred by me traveling to and from, or while located in _____.

_____   _____
Witness   (Signature)

Dated: _____

    City of _____

    State of _____

LIFELINE EARTHQUAKE INVESTIGATIONS: PAST AND FUTURE

Investigation of damage and system response after earthquakes has been one of the primary sources of knowledge for improving building codes, seismic design of equipment, and system response to damaging earthquakes. Lifeline systems include power, communication, water, sewage, transportation and liquid and gaseous fuels systems. The investigation of lifeline systems poses special problems. Lifeline equipment and structures are often highly specialized, the effect of damage has to be evaluated in terms of how it impacts system response and lifelines are highly interdependent.

After the Alaskan earthquake of 1964 the National Research Council (NRC) of the National Academy of Sciences began investigating earthquakes and other natural disasters. Since 1971 the Earthquake Engineering Research Institute (EERI) has also sent reconnaissance teams to evaluate the impact of significant earthquakes throughout the world. In recent years, the Earthquake Investigations Committee of the Technical Council on Lifeline Earthquake Engineering, ASCE, has provided an investigator to EERI and NRC reconnaissance teams to investigate lifeline damage. Over the years teams have investigated earthquakes in Japan, Mexico, Chile, Italy, Armenia, the Philippines and many other countries as well as numerous US earthquakes.

Early investigations emphasized the documentation of damage. Recently, increased emphasis has been placed on determining the causes of damage and its impact and documenting equipment which has performed well. This expanded scope has put an impossible burden on the lifeline investigator. The individual would have to investigate each of the lifelines mentioned above in great detail. For each lifeline he would have to be familiar with the equipment, facilities, and the operation of the system.

To improve post-earthquake investigations of lifelines more investigators are needed to gather information and each investigator must have broader knowledge of each of the lifelines. In addition, investigators must be familiar with past earthquake damage in order to identify and evaluate types of damage not previously observed.

To meet the need for more and better trained investigators, the Earthquake Investigations Committee has conducted workshops to train investigators.

The demands on an investigator are highly dependent on the earthquake and its location. Investigations within the U.S. may take two or three days, while those in foreign countries typically require over a week in the field. Individuals must be able to travel on short notice. Upon returning, a detailed report and a paper for publication must be prepared describing observations. In addition, some individuals may be requested to make a presentation at a briefing session. The costs associated with the investigation such as transportation, food and lodging will have to be borne primarily by the individual or his employer. It should be noted that while no investigator has ever been seriously injured, there are inconveniences and risks in going to an area that has just had a damaging earthquake.

All investigators with whom the author has spoken have said that the earthquake investigation experience was professionally valuable to them and their employers, as well as personally rewarding.

Summary of objectives of post-earthquake investigator training.

1. General topics: what to bring, what to do before you depart, what to do when you get there, how to conduct an investigation, and how to record data.

2. An overview of each lifeline system including a discussion of all facilities and equipment. The role of equipment and facilities in the operation of the system will be explained.

3. The physical characteristics of each lifeline's facilities and equipment will be discussed with emphasis on those that have had seismic problems.

4. A detailed review of past earthquake performance of equipment and facilities for each lifeline.

5. Special investigation methods applicable to each lifeline.

The first workshop will emphasize power systems, water, and sewage systems.

THE POST-EARTHQUAKE INVESTIGATION PROCESS

Determining If a Reconnaissance Team Will Be Dispatched

Formal post-earthquake investigations have historically been conducted under the auspices of two organizations, Earthquake Engineering Research Institute (EERI) and the National Research Council (NRC) of the National Academy of Sciences (NAS). Since 1974 the Earthquake Investigations Committee, TCLEE has provided a lifeline investigator for the reconnaissance team. At the present time, when there is a significant earthquake anywhere in the world, the Learning from Earthquakes Project Manager of EERI and the Disasters Investigations Coordinator of NRC determine if the earthquake warrants dispatching a reconnaissance team. Factors which influence this decision include the size the the earthquake, the character of the impacted area (The key questions is, can useful information be derived from an investigation?), and the accessibility of the damaged area. Information from personal contacts with individuals near the damaged area is often used to determine if a team should be dispatched, and for small earthquakes an individual on the scene may conduct the investigation. EERI and NAS may decide that the earthquake does not warrant dispatching a reconnaissance team. The Earthquake Investigations Committee may still investigate the earthquake if it appears appropriate from the lifelines perspective. For examples, the Earthquake Investigations Committee took the leading role in organizing an investigation following the Tejon Ranch earthquake of June 10, 1988 and the Philippine earthquake of July 16, 1990

Selecting Team Members for Lifeline Investigation

The selection procedure is somewhat different for U.S. and foreign earthquakes. With the availability of more trained lifeline investigators, and the realization that more detailed lifeline information has to be collected than in the past, it is anticipated that the Earthquake Investigations Committee will place about five people in the field for a significant foreign earthquake. Team members will be selected so that expertise for most lifelines will be represented, with a weighting towards those lifelines that are known to have suffered extensive damage. Language skills of the investigator will also play an important role in the selection. Finally, it would be desirable to have at least one individual who has field experience in conducting an investigation. It is anticipated that the lifeline team will be part of the EERI or NAS team so selection will have to be coordinated with these groups. Financial support for investigations may be limited so that some of all of an investigators expenses may not be reimbursed.

For U.S. earthquakes it will be possible to get more people in the field because transportation cost would be lower and travel times would be less. It would also provide an opportunity to give people field experience and training, and investigation methods could be evaluated and improved. For a significant earthquake, it should be possible to have as many individuals on the team as are interested in participating.

Earthquake Investigation

The earthquake investigation would be conducted using the guidance provided by the Training Guide and according to the dictates of the situation.

Preliminary Report

A preliminary report should be submitted by each investigator within **one week** of returning from the earthquake. This report should not be more than about page long, list the major observations, and provide an estimate of the length of the individual's contribution to the final report. The preliminary report serves three functions. First, it provides information for planning a debriefing session. Second, it provides a basis for estimating the scope of the papers that are to be published, and finally, it provides information for brief summary reports that will be published in ASCE NEWS and the EERI NEWSLETTER. These summary reports are usually limited to several column-inches (ASCE NEWS) or a few pages (EERI NEWSLETTER).

Earthquake Investigation Debriefing (Damage Observations)

For some earthquakes where the technical community has a great interest in the earthquake and its effects, one or more debriefing sessions will be held. For example after the Whittier Narrows Earthquake of 1987, EERI had debriefing sessions in Los Angeles, San Francisco, Seattle and Washington D.C. Because of the large number of topics that have to be discussed, one individual will present all the observations for lifelines, due to time limitations. If more than one debriefing is given, different investigators may make the presentation depending on their availability and their location relative to the session. These sessions are usually held about a month after the earthquake. The expenses of speakers for these session are covered by admission charges to the debriefing. If income exceeds the expenses, a small honorarium may be paid to speakers.

Paper for Publication

Since the introduction of EARTHQUAKE SPECTRA by EERI, reconnaissance reports have been published in it. For larger earthquakes, an entire issue is devoted to the earthquake. For smaller earthquakes, the report may be issued as a separate reconnaissance report. Papers may also be prepared for an ASCE Journal, TCLEE monograph series and for trade publications. The details of how the published report is prepared may depend on the situation in which the data are gathered. For some individuals there are important reasons for them to be identified as a principal author of a publication. However, if a team gathers the data and prepares the report, individual authorship may not be possible.

Detailed Reconnaissance Report

In addition to the paper that is prepared and published, most investigations generate data that for one reason or another is inappropriate to include in a formal publication, but which is of value to the professional community. Examples are manufacturers' model numbers of damaged equipment, or very speculative views of what might have contributed to a failure. In addition, names and contacts made during the investigation may be useful for follow-up studies Thus, the detailed reconnaissance report would consist of the published report and supplemental data, names and addresses of contacts, speculations on factors contributing to the damage, sketches, and other material that may be of use to future investigators.

Slides for the Master File

   Selected slides that illustrate the different types of damage, factors that may have contributed to the damage and solution to mitigate earthquake effects will be collected and made available for general use. Each slide should be annotated.

Earthquake Investigation Procedure Evaluation Report (Observations on the data collection process.)

   The current state-of-the-art of post-earthquake lifeline investigations is just emerging so that this Guide should see significant improvements as field experience is gained. It is important that the lessons learned about the process of collecting detailed information be documented in a report.

Analysis of the Post-earthquake Investigation Process

   At the annual meeting of the Earthquake Investigations Committee following each earthquake investigation, there will be a discussion of the lessons learned on the process of gathering earthquake damage and system response data. It is anticipated that the investigators would discuss their Earthquake Investigation Procedure Evaluation Report and changes would be collected for updating the Guide.

2. General Pre-Departure Preparations

   Personal Needs
   Materials for the Investigations

2. GENERAL PRE-DEPARTURE PREPARATIONS

Some material and information should be gathered and made ready before embarking on an investigation. These preparations will be of particular importance for an investigation outside of the U.S. Also, these recommendations are for a <u>worst-case</u> scenario for an investigation of a major earthquake outside of the U.S. For a U.S. investigation, some of the materials and comments will not be necessary or applicable. Since some materials will age, such as batteries and film, they should be gathered at the time of the earthquake before departure. During the investigation you will be on the move so that luggage should be kept light and compact. All materials should be able to fit into a carry-on bag and shoulder bag.

Your personal physician should be consulted for advice about medical precautions you should take. Some treatments have risks associated with them so overall risk should be considered. Some hospitals have travel medicine sections that specialize in precautions to take when traveling. It would be desirable to locate one in advance so that they can be consulted when travel location is known.

You should also locate a source that has a large selection of travel maps. Very often maps cannot be obtained when you arrive in the country. The map brought from home also is more likely to be in English. You will probably still want to obtain more detailed maps when you arrive at the scene.

Establish contacts with the trade or professional organization for the lifeline for which you are a specialist. They may be able to provide you with local contacts and some background information.

PERSONAL NEED

Travel Documents

Passport (You should have a valid passport at all times.)

Visa (If a visa is required it will have to be obtained after the earthquake.)

Immunization (vaccination) certificates (See Health Needs below.)

Backup copies of the above documents. Extra passport pictures are some times needed.

Business cards are very useful and in some countries they are exchanged on all introductions.

Money

Cash, travelers checks and credit cards should be taken. It is suggested that more cash be taken than one would normally carry on a vacation trip as there may be reluctance to take credit cards due to the disruption of banking services. In some places travelers checks denominated in the local currency simplifies cashing.

Before departing, some money should be converted into the currency of your destination.

Passport Case

A cloth passport case than can hold your passport and cash that has a string so that it can be worn around around the neck can be very useful.

Health Needs

Personal medication as required. Persons with serious health problems should not undertake an investigation in a seriously damaged area.

Immunization certificates are required for entry into some countries. While a travel agent can provide you with information as to what shots are required for a given area, it is highly recommended that you consult with your own physician or with a physician specializing in travel medicine. Most large hospitals have a travel medicine clinic, which is likely to be the best source of information about potential risks outside of the U.S. They can provide advice concerning required and optional immunizations. They will also be aware of endemic diseases which may normally pose little threat to the traveler, but because of the breakdown in sanitation following an earthquake may be a serious health problem. Specialized departments will usually have computer terminals directly connected to the Center for Disease Control so current information would be available.

The fact that you will be in an earthquake damaged area means that water supply and other conditions may not be up to their usual standards. While the earthquakes that are usually investigated have impacted large metropolitan areas, impacted lifeline facilities often are outside of urban areas. The situation should be discussed with medical personnel as well as any personal health problems that should be taken into consideration.

Health Information for International Travel (HHS Publication No. (CDC)xx-xxxx) published by the U.S. Public Health Service is updated periodically and a current copy should be consulted. The publication is available from

> Superintendent of Documents
> US Government Printing Office
> Washington D.C. 20402

The Foreign Quarantine Division of the Center for Disease Control in Atlanta also publishes a weekly "Blue Sheet" indicating changes in the requirements of countries for required vaccinations. Conditions brought about by the earthquake may cause a change in the vaccination requirements.

Water quality will often be suspect. It is suggested that two one-quart, plastic water bottles be taken.

Water purification pills should be brought to cover your needs for the entire trip. Plan for at least two quarts a day. A water filter may be useful.

It is not uncommon for investigators to develop diarrhea. You should consult with your physician for the recommended medication and bring a supply .

Face masks are useful for protection against dust or odors. Disposable dust masks can be purchased at most hardware stores.

Work gloves to protect the hands from dirt and abrasions are desirable.

If you do not have a hard hat, you should make arrangement to borrow one.

If you require glasses, a spare pair and elastic retainers such as are used for sports should be taken. Sun glasses should be considered.

A small first aid kit containing the following as appropriate. Band-aids, antiseptic, sun screen, insect repellent, aspirin.

Some high energy snack foods should be brought along as some days it may be difficult to get food during the day. In extreme situations, provision for all food needs would have to be made.

Ear plugs can be very useful. They can not only afford you better sleep if you get a room mate who snores, but if a helicopter flight can be arranged or if non-commercial transportation is used (A cargo plane was used to get to Armenia) ear plugs may be necessary to prevent loss of hearing.

Personal Clothing

Note that much of your time will be spent outdoors and in unheated structures. Sturdy shoes or boots are important. Very often it will be necessary to climb within a structure to get assess to damaged areas and equipment so that a parka or wind breaker rather than a trench coat is more practical. Depending on local weather conditions, raingear and/or cold weather clothing (wool sweaters, hat, stockings, etc.) may be required. Even in hot climates a light weight long sleeve shirt should be taken. It can provide protection from the sun, and many electric power substations require that arms be protected (with cotton shirt rather than one made of synthetic material) to enter the substation.

Tips on cold weather. These tips will have to be viewed as a worst case situation in which one may have to live out of doors and have little support in a winter environment. Very extreme conditions were encountered after the Armenian earthquake when staying in the damaged area. The suggestions are based on mountaineering experience and are personal views that others may have different ideas about.

Keeping warm is more than just keeping comfortable, your health is at risk. While frostbite of extremities is one concern, improper or inadequate clothing can lead to hypothermia and death after a couple hours exposure.

While a down parka is the warmest garment for its weight and packed size, it is of no use if it gets wet, so if rain is a possibility several layers of wool should be used for warmth. A Gortex shell provides good wind protection and it the garment is properly made and not old can provide good rain protection. At the moderate levels of activity encountered during an earthquake investigation it will breathe so that garments under the parka will remain relatively dry as compared to a waterproof parka. Rain pants should be taken as they provide wind protection as well as rain protection. A rain poncho is ill suited for an earthquake investigation, although it can serve as a ground cloth or bivouac sack.

A wool hat that can cover the ears is important as is an ear band (a wool band that covers the ears) should a hard hat have to be warn. Two pairs of mittens should be taken, one a very light pair and a heavier pair. Mittens are warmer than gloves and most things, such adjusting camera, can be done with them on. One can buy a mitten that allows the fingers to be slipped out to do fine work. Another reason for taking two pairs is redundancy as it is easy to loose a mitten. A wool scarf is useful as it can be used to protect your neck, a major point of heat loss, and also serve to protect ears or hands should the need arise.

Footwear is vital. In general two pairs of wool stockings should be warn, a light pair under a heavier pair. The very heavy hunter type stockings are difficult to dry out if they get wet. The use of two stockings will usually prevent chafing. At least three changes of stockings should be taken. It can be useful to have a large safety pin to attach a damp pair of stockings under your coat to dry them out. For wet weather or if you will be in water a hunting shoe, that is a rubber bottom with leather or rubber tops provides good protection. These should have removable felt liners. A spare set of felt innersoles can be useful. The bottoms of footwear should be cleated to provide good footing on snow or mud.

Emergency Food

Two types of emergency food should be taken. The first type which may be useful in less severe situations is snack food to substitute for lunches that may not be available while in the field during the day. The second type would be food that would serve as a substitute for normal food sources. The amount of food would depend on the situation that might arise. This food should have several characteristics. It must be high energy which means that it should be oil based, it should keep unrefrigerated for an appropriate period, it should not require heating or cooking, and you should find it acceptable as a diet or a several days. Examples would be peanut butter, cheese, and sausage along with firm breads that will keep.

MATERIALS FOR THE INVESTIGATION

Cameras

Two cameras are suggested. You should be very familiar with the operation of the cameras. Lifeline equipment is often inside structures that are poorly lighted. For this reason a small automatic camera is very useful since it will have an auto-focus, auto-exposure, with built-in flash and the camera is relatively light. While a single lens reflex (SLR) with a zoom lens provides more flexibility in framing pictures, in poorly lighted locations there is often insufficient light to focus the camera. Their use is very time consuming and they are bulky. The newer, top of the line automatic camera also have limited wide angle, telephoto and macro capabilities built in to the camera. One disadvantage of theses types of cameras is that the built in flash units are not very powerful so that pictures taken at any distance are often poorly lighted. Powerful, flash attachments do a better job.

A macro lens on at least one camera is very important for many situations. Very often for lifelines, restoration starts within hours of the earthquake so that damage is cleared before the investigation team can get to the site. In foreign earthquakes it may be a week before the team is in the field. By that time pictures taken by local personnel of damage that has already been repaired are often available at the sites. With a macro lens good copies can be obtained from these photographs. If a slide copier attachment is available for you camera, it should be taken. While copies of slide and pictures can be requested, they are seldom forthcoming.

In general, only one camera is used, but, because pictures are so important for documenting damage, the second camera serves primarily as a spare should the first malfunction or get damaged while climbing around a structure. Enough high speed color film for the entire trip should be taken as film may not be available locally and it is typically more expensive. The amount of film will depend on the individual's style, but a minimum of 3 rolls per day should be taken, more for a shorter trip, less for a longer trip. Depending on the magnitude and location of the earthquake, film supplies may be replenished in the field. A film travel bag should be considered to prevent an out-of-adjustment airport X-ray machine from fogging your film. Alternatively, you may carry film and request a personal search. Automatic cameras consume a lot of batteries, since the flash is often used. You should bring about one change of batteries for every three rolls of film, although this will be dependent on the type of camera used. If the weather is cold, take 50% more batteries. In cold weather one set of batteries should be kept in an internal pocket so that they will be warm when they are needed. In cold dry weather the film should be rewound slowly to reduce the possibly of generating static electricity that can damage pictures. Many non-automatic cameras require a battery for a light meter so a new battery installed before departure or spare battery should also be taken as they may not be available in the field. SLR cameras typically require a separate flash unit.

Tape Recorder

One of the very small tape recorders is very useful. A voice activated type is handy but it should have a manual over-ride for very noisy environments. This can be used to simplify record keeping, and recording interviews. Two notes of caution. Unless you are used to recording and you have some one to transcribe your records, this can be an unpleasant, time consuming task after the earthquake (the recorder can be useful for preparing a written summary at the end of each day and serve as a more complete record if it has to be referred to.) A potential problem associated with recording interviews is that some people will be very self-conscious if they are going to be recorded and may be less forthcoming. All tape and battery needs for the entire trip should be brought with you. Tape usage depends upon individual preference but 2 hours per day should be ample.

Note Pads

> A steno pad with its hard cover simplifies taking notes and it can be put into a coat pocket. A small note pad is also useful since it can fit in a shirt pocket. A clipboard should be considered.

Maps

> You will need good maps for the region that you are going to. Maps can be obtained from USGS, specialty travel book stores, and from some university libraries that maintain special map collections.

Hand Held Copier

> In many countries the availability of maps is very limited. Also, when damage is severe, detailed city maps are desirable and they must usually be obtained after you are in the field. Local authorities may not have copies or be able to get copies made. For this special situation a single hand held copier for team use would be very helpful.

List of Contacts

> A list of facilities and contacts (including phone numbers) that has been gathered prior to departure.

Aids

> Tape measure, machinist scale, micrometer, magnetic compass, flashlight with spare batteries, magnifying glass, binoculars or monocular, foreign language dictionary, ruler for pictures (See Appendix A), north arrow for pictures (See Appendix A)

Field Guide

> The field guide contains check list and intensity scales.

Small Shoulder Bag

> A small shoulder bag or "fanny pack" (type used by skiers) to carry your needs while you are away from transportation vehicle is important. This would carry several rolls of film, camera attachments, and aids needed to conduct the investigations such a scale, compass, etc. This bag should be easy to manage while climbing in and out of vehicle and around damage sites. The "fanny pack" is recommended as it is less cumbersome than a shoulder bag.

Large Shoulder Bag

    A large shoulder bad is vital for daily needs. It must be large enough to carry all of your materials except for extra clothes and some of your film supplies. A large multi-compartment "gym" bag that can hold your extra camera, film, wind breaker, rain shell, etc. should serve the purpose.

Suit Case

    A small suit case should be used to pack materials. One with a shoulder strap can be useful.

3. Post-Earthquake/Pre-Departure Preparations

    Post-Earthquake - Pre-departure Tasks
    General Pre-Investigation Tasks for Significant U.S.
        Earthquakes: Team Leader Tasks
    Departure Check List

3. POST-EARTHQUAKE/PRE-DEPARTURE PREPARATIONS

There are post-earthquake - pre-departure tasks for the individual investigator and for the team leader.

POST-EARTHQUAKE - PRE-DEPARTURE TASK

Once you are informed that you will be a member of a reconnaissance team several pre-departure tasks should be completed.

Make sure the following information is understood by team members. Determine where the team will meet in the U.S. before leaving the country and where the team expects to meet in the destination country. This is important since the team may be assembled from around the country and due to a missed connection you may not make the U.S. meeting place. Determine a central contact place where you can call to pass information on to the team leader. Since he may be traveling, contact would have to be made through this contact point.

Check with health authorities to determine what vaccinations are recommended and if verification of immunization is required. They can also advise if any special health hazards are expected in the area of the investigation. They may not be aware of special circumstances associated with the earthquake.

Check your passport and request a visa if required. To expedite the visa process, a trip to a consulate may be needed.

Get maps for the areas impacted by the earthquake.

Buy film, recording tape, and batteries required for the entire trip (special camera battery, batteries for automatic cameras, flashlight batteries, tape recorder batteries). A sequence number should be scratched on each film case and put on the outside of each film canister. Once in the earthquake zone, these materials may not be available.

Contact friends, colleagues, professional organizations for information on the facilities of interest in the area and names and phone numbers of contacts within the country. University professors will often have former students in the country, who can be invaluable in helping with an investigation. They will know the customs, language and have their own additional contacts. Language is a very big problem, even if you have a team member who is fluent in the language. Technical terms are often different and good translation will require someone who has used the technical terms within the country.

Set up a network to use to relay information back in the U.S. Additional significant aftershocks may occur while you are investigating the earthquake and your family will be concerned. It may be possible to make international calls from the earthquake zone, but they can be very time consuming. If possible make arrangements so that a single contact in the U.S. can relay messages.

Gather materials and check Departure Check List to see if you have forgotten anything. (See Appendix A.)

GENERAL PRE-DEPARTURE TASKS FOR SIGNIFICANT U.S. EARTHQUAKES: TEAM LEADER TASKS

- Revise Pre-Investigation requests for information to present situation.

- Identify individuals to distribute information requests.

- Get newspaper subscription to local papers and set up readers and communication links.

- Establish lifeline coordinators for each investigation area. The following areas should be investigated, although the list would have to be tailored to the circumstances of the particular earthquake.

  Power
  Communications
   Telephone Communications
   Emergency Response communications including PSAP, Ham
   Cellular Telephones
   Paging Systems, including automatic security and fire systems
   Newspapers and Broadcast media including emergency broadcast system
  Gas Systems including local and transmission natural gas, and liquid fuels
  Water
  Sewage systems including local gathering systems and regional treatment facilities
  Transportation
   Highways
   Bridges
   Railways
   Airports
   Ports
   Regional Transportation Authority

DEPARTURE CHECK LIST

___ Passport, Visa, immunization (vaccination) certifications, and copies
___ Business cards
___ Cash, credit cards, local currency for destination
___ Personal medication, small first aide kit
___ Spare and sun glasses and glasses elastic safety strap
___ Camera, film and spare batteries
___ Spare camera
___ List of contacts and information on designation
___ US and destination contact points and phone numbers
___ Water bottles, water purification pills, water filter
___ Diarrhea medication
___ Dust mask
___ Hard hat
___ Gloves
___ High energy snack foods
___ Personal clothing
___ Tape recorder, tape and batteries
___ Note pads, clipboard
___ Maps
___ Tape measure, scale, micrometer
___ Magnetic compass
___ Magnifying glass
___ Flashlight with spare batteries
___ Foreign language dictionary
___ Ruler and north arrow for pictures
___ Field Guide
___ Small Shoulder bag, large shoulder bag, suite case, passport case

4. Introduction to Earthquakes and Their Effects

    Ground Vibrations
    Soil Liquefaction
    Earthquake Induced Landslides
    Subsidence
    Ground Faulting
    Earthquake-Induced Water Waves (Tsunami)
    Seismic Hazard Maps
    Commonly Used Terms
    Intensity Scales
        Modified Mercalli Intensity Scale
        Rossi-Forel Scale
        MKS Intensity Scale

4. INTRODUCTION TO EARTHQUAKES AND THEIR EFFECTS

The understanding of the causes of earthquakes has been revolutionized in the last 50 years and in many regions, such as the eastern US, much remains to be learned. It is a surprise to many that the largest and most widely felt earthquakes to effect the US had their origin in the Midwest, not California. The release of pent-up strain energy in the form of seismic waves have both direct and secondary effects that impact lifelines. The damaging effects of earthquakes can be felt hundreds of miles from their origin and can be significantly influenced by local topography and soil conditions.

Knowledge of the effects of earthquakes that impact lifelines has been gained from observing earthquakes that have occurred in many parts of the world. The significance that any effect has to a particular region depends on the characteristics that prevail there. Each of the following sections will discuss one of the major effects exhibited by earthquakes. There is an emphasis on those effects which are most important in the eastern US. The last section will define some terms often encountered when dealing with earthquakes. This brief review is primarily to define terms and establish a minimum common denominate for communication.

GROUND VIBRATION

When an earthquake occurs seismic energy radiates away from the source in the form of seismic waves. These waves exhibit themselves by vibration of the ground. The vibration of the ground will induce vibration in the structures and equipment resting on the ground. Equipment mounted in structures will experience the motion induced in the structure by the ground shaking. In general the magnitude of the ground shaking decreases as the distance from the source increases, however, local soil conditions can amplify vibration levels by a factor of three or more.

Earthquake excitations can be characterized by the amplitude of the shaking and its frequency content. The frequency of the high energy content of earthquake ground motions coincides with the frequency of a significant portion of lifeline facilities and equipment. The effects of induced vibration is the major cause of lifeline damage.

SOIL LIQUEFACTION

Under certain condition, when soils experience vibrations, the phenomenon of soil liquefaction can be observed. When soil liquefies, it loses its shear strength. Liquefied soil has been observed to flow on 1 % grades and surface supported structures have settled several feet below grade; buried tanks that were lighter than the material in which they were buried have floated to the surface. While there is a tendency to emphasize vertical motions associated with soil liquefaction, extensive horizontal or lateral spreading has been observed.

Factors contributing to liquefaction are the amplitude and duration of shaking, height of the water table, the soil density and the granular character of the soil. These conditions are common adjacent to rivers and lakes, common sites for power facilities. Conditions in the East and Midwest US suggest that more liquefaction can be expected than in California.

Earthquake Induced Landslides

There are many regions in which the earthquake induced shaking triggers landslides. The topography and soil conditions are the primary control variables but should the earthquake occur during a rainy season when soils are saturated, the situation can be aggravated. Slides can cause excessive deformations in the ground or the motion of the soil may sweep away structures and equipment in it path.

Subsidence

Under certain conditions, earthquake-induced vibrations may cause extensive ground settling. In past earthquakes this has caused flooding and differential settlement with attendant severe structural loads. Large cracks in the ground can appear as a result of subsidence due to differential settlement and should not be confused with earthquake faulting.

Ground Faulting

Faults are fracture planes where there is relative motion between the rock on each side of the fracture. In settings with multiple fractures the area is referred to as a fault zone. Thus, anything spanning the fault, such as buried pipe or cable or a structure, can experience severe deformations and loads. Depending on the earthquake, the motion across the fault can be both horizontal and/or vertical with displacements of 20 feet. In some earthquakes, particularly in the eastern US where there are deep alluvial deposits, the faulting may not extend to the ground surface so faulting is not observed.

Earthquake-Induced Water Waves (Tsunami)

Earthquakes occurring off shore that have vertical components can generate large, long period waves. They can have significant impact of coastal structures as they can have on shore run up as large a 6 meters high. For small on-shore grade, the run up can extend large distances inland.

Seismic Hazard Maps

Various seismic hazard maps are available, however, most detailed maps are limited to California. USGS is in the process of developing slope and soil stability maps for other regions of the country. National code maps, such as the Uniform Building (UBC) Code Risk map, are of little value for earthquake investigations.

Commonly Used Terms

A. Epicenter - The point on the earths surface just above the point where the earthquake faulting started.

B. Richter Magnitude - A common measure of the size of an earthquake derived from seismograms obtained from the earthquake and related to the energy released in the earthquake. Near the epicenter a M = 6 (1971 San Fernando M = 6.4) can cause significant damage, M = 7 is a major earthquake, and M = 8 is a great earthquake (1906 San Francisco M = 8.3). It should be noted that an increase in the Richter Magnitude of 1 correspond to about a 30 fold increase in the energy released.

C. Modified Mercalli and Rossi-Forel Intensity Scale - Scales based on observed damage which quantifies the local intensity of ground shaking. See Appendix A.

Modified Mercalli Intensity Scale

I   Note Felt. Marginal and Long-period effects of large earthquakes.
II Felt by persons at rest, on upper floors, or in favorable places.
III Felt indoors. Hanging objects swing. Vibration like passing of light trucks. Duration estimated. May not be recognized as an earthquake.
IV Hanging objects swing. Vibration like passing of heavy trucks; or sensation of a jolt like a ball striking the walls. Standing motor cars rock. Windows, dishes, doors   rattle. Glasses clink. Crockery clashes. In the upper range of IV wooden walls and frames creak.
V   Felt outdoors; direction estimated. Sleepers wakened. Liquids disturbed, some spilled. Small unstable objects displaced or upset. Door swing, close and open. Shutters, pictures move, Pendulum clocks stop, start, change rate.
VI Felt by all. Many frightened and run outdoors. Persons walk unsteadily. Windows, dishes, glassware broken, knickknacks, books, etc. fall off shelves. Pictures fall off walls. Furniture moved or overturned. Weak plaster and masonry D cracked. Small bells ring in churches and schools. Trees, bushes visibly shaken or heard to rustle.
VII Difficult to stand. Noticed by drivers of motor cars. Hanging objects quiver. Furniture broken. Damage to masonry D, including cracks. Weak chimneys broken at roof line. Fall of plaster, loose bricks, stones, tiles cornices (also unbraced parapets and architectural ornaments). Some cracks in masonry C. Waves on ponds; water turbid with mud. Small slides and caving in along sand or gravel banks. Large bells ring. Concrete irrigation ditches damaged.
VIII Steering of motor car affected. Damage to masonry C; partial collapse. Some damage to masonry B; none to masonry A. Fall of stucco and some masonry walls. Twisting, fall of chimneys, factory stacks, monuments, towers, elevated tanks. Frame houses moved on foundations if not bolted down; loose panel walls thrown out. Decayed piling broken off. Branches broken from trees. Changes in flow or temperature of springs and wells. Cracks in wet ground and of steep slopes.
IX General Panic. Masonry D destroyed; masonry B seriously damaged. (General damage to foundations.) Frame structures, if not bolted, shifted off foundations. Frames racked. Serious damage to reservoirs. Underground pipes broken. Conspicuous cracks in ground. In alluviated areas sand and mud ejected, earthquakes fountains, sand craters.
X   Most masonry and frame structures destroyed with their foundations. Some well-built wooden structures and bridges destroyed. Serious damage to dams, dikes, embankments. Large landslides. Water thrown on banks in canals, rivers, lakes, etc. Sand and mud shift horizontally on beaches and flat land. Rails bent slightly.
XI Rails bent greatly. Underground pipelines completely out of service.
XII Damage nearly total. Large rock masses displaced. Lines of sight and level distorted. Objects thrown into air.

After Richter, C.F. Elementary Seismology.

Note: To avoid ambiguity, the quality of masonry, brick, or other material is specified by   the following system. (This has no connection with the conventional classes A, B,   and C construction.)

Masonry A. Good workmanship, mortar, and design; reinforced, especially laterally, and bound together by using steel, concrete, etc.; designed to resist lateral forces.

Masonry B. Good workmanship and mortar; reinforced, but not designed to resist lateral forces.

Masonry C. Ordinary workmanship and mortar; no extreme weaknesses, like failing to tie in at corners, but neither reinforced nor designed to resist horizontal forces.

Masonry D. Weak materials, such as adobe; poor mortar; low standards of workmanship; weak horizontally.

Rossi-Forel Scale (After Richter, 1956)

I   Microseismic shock. Recorded by a single seismograph or by seismographs of the same model but not by several seismographs of different kinds: the shock felt by an experienced observer.
II Extremely feeble shock. Recorded by several seismographs of different kinds; felt by a small number of persons at rest.
III Very feeble shock. Felt by several persons at rest; strong enough for the direction or duration to be appreciated.
IV Feeble shock. Felt by persons in motion; disturbance of movable objects, doors, windows, cracking of ceilings.
V   Shock of moderate intensity. Felt generally by everyone; disturbance of furniture, beds, etc., ringing of some bells.
VI Fairly strong shock. General awakening of those asleep; general ringing of bells; oscillation of chandeliers; stopping of clocks; visible agitation of trees and shrubs; some startled persons leaving their dwellings.
VII Strong shock. Overthrow of movable objects; fall of plaster; ringing of church bells; general panic, without damage to buildings.
VIII Very strong shock. Fall of chimneys; cracks in the walls of buildings.
IX Extremely strong shock. Partial or total destruction of some buildings.
X   Shock of extreme intensity. Great disaster; ruins; disturbance of the strata, fissures in the ground, rock falls from mountains.

MKS Intensity Scale (From World Data Center A for Solid Earth Geophysics, Report SE-20 1979.)

I   Not noticeable a). The intensity of the vibration is below the limit of sensibility; the tremor is detected and recorded by seismographs only.

II Scarcely noticeable (very slight) a). Vibration is felt only by people at rest in houses, especially on upper floors of buildings.

III Weak, partially observed only. a). The earthquake is felt indoors by a few people, outdoors only in favorable circumstances. The vibration is like that due to the passing of a light truck. Attentive observers notice a slight swinging of hanging objects, somewhat more heavily on upper floors.

IV Widely observed. a). The earthquake is felt indoors by many people, outdoors by a few. Here and there people awake, but no one is frightened. The vibration is like that due to the passing of a heavily loaded truck. Windows, doors and dishes rattle. Floors and walls creak. Furniture begins to shake. Hanging objects swing slightly. Liquids in open vessels are slightly disturbed. In standing motor cars the shock is noticeable.

V   Awakening a). The earthquake is felt indoors by all, outdoors by many. Many sleeping people awake. A few run outdoors. Animals become uneasy. Buildings tremble throughout. Hanging objects swing considerably. Pictures knock against walls or swing out of place. Occasionally pendulum clocks stop. A few unstable objects may be overturned or shifted. Open doors and windows are thrust open and slam back again. Liquids spill in small amounts from well-filled open containers. The sensation of vibration is like that due to a heavy object falling inside the building. b). Slight damage of Grade 1 in buildings of Type A is possible. c). Sometimes changes in flow of springs.

VI Frightening a). Felt by most people indoors and outdoors. Many people frightened and run outdoors. A few persons lose their balance. Domestic animals run out of their stalls. In a few instances, dishes and glassware may break, books fall down. Heavy furniture may possibly move and small steel bells may ring. b). Damage of Grade 1 is sustained in single buildings of Type B, and in many of Type A. Damage in a few buildings of Type A is of Grade 2. c). In a few cases crakes up to widths of 1 cm possible in wet ground; in mountains occasional landslides; change on flow of springs and in the level of well-water is observed.

VII Damage to buildings. a). Most people are frightened and run outdoors. Many find it difficult to stand. The vibration is noticed by persons driving motor cars. Large bells ring. b). In many buildings of Type C damage of Grade 1 is caused; in many buildings of Type B damage is of Grade 2. Many buildings of Type A suffer damage of Grade 3, a few of Grade 4. In single instances landslides of roadway on steep slope, cracks in roads, seams of pipelines damaged, cracks in stone walls. c).Waves are formed on water, and water is made turbid by mud stirred up. Water levels in wells change , and the flow of springs change. In a few cases dry springs have their flow restored and existing springs stop flowing. In isolated instances parts of sandy or gravelly banks slip off.

VIII Destruction of buildings. a) Fright and panic; also persons driving motor cars are disturbed. Here and there branches of trees break off. Even heavy furniture moves and partly overturns. Hanging lamps are in part damaged. b) Many buildings of Type C suffer damage of Grade 2, a few of Grade 3. Many buildings of Type B suffer damage of Grade 3, and many buildings of Type A suffer damage of Grade 4. Occasional breakage of pipe seams. Memorials and monuments more and twist. Tombstones overturn. Stone walls collapse. c) Small landslide in hollows and on banked roads on steep slopes; cracks in ground up to widths of several centimeters. Water in lakes becomes turbid. New reservoirs come into existence. Dry wells refill and existing wells become dry. In many cases changes in flow and level of water.

IX General Damage to buildings. a). General panic; considerable damage to furniture. Animals run to and fro in confusion and cry. b) Many buildings of Type C suffer damage of Grade 3, a few of Grade 4. Many buildings of Type B show damage of Grade 4, a few of Grade 5. Many buildings of Type A suffer damage of Grade 5. Monuments and columns fall. Considerable damage to reservoirs; underground pipes partly broken. In individual cases railway lines are bent and roadways damaged. c). On flat land overflow of water, sand and mud is often observed. Ground cracks to widths of up to 10 cm. on slopes and river banks more than 10 cm; furthermore a large number of slight cracks in ground; falls of rock, many landslides and earth flows; large waves on water. Dry wells renew their flow and existing wells dry up.

X General destruction of buildings. b). many buildings of Type C suffer damage of Grade 4, a few of Grade 5; critical damage to dams and dikes, and severe damage to bridges. Railway lines are bent slightly. Underground pipes are broken or bent. Road paving and asphalt show waves. c) In ground, cracks up to widths of several tens of centimeters, sometimes up to a meter. Broad fissures occur parallel to water courses. Loose ground slides from steep slopes. From river banks and steep coasts considerable landslides are possible. In coastal areas displacement of sand and mud; change of water level in wells; water from canals, lakes rivers, etc. thrown on land. New lakes occur.

XI Catastrophe b) Severe damage even to well-built buildings, bridges, water dams and railway lines; highways become useless; underground pipes destroyed. c) Ground considerably distorted by broad cracks and fissures, as well as by movement in horizontal and vertical directions; numerous landslips and falls of rocks. The intensity of the earthquake requires to be investigated specially.

XII Landscape changes. b) Practically all structures and below ground are greatly damaged or destroyed. c) The surface of the ground is radically changed, considerable ground cracks with extensive vertical and horizontal movements are observed. Fall of rock and slumping of river banks over wide areas; lakes are dammed; waterfalls appear, and rivers are deflected. The intensity of the earthquake requires to be investigated specially.

Types of Structures

Structure A   Building in field-stone, rural structure, adobe houses, clay houses.
Structure B   Ordinary brick buildings, buildings of the large block and prefabricate type, half timbered structures, buildings in natural hewn stone.
Structure C   Reinforced buildings, well-built wooden structures.

Definition of quantity: Single few - about 5%; Many - about 50%; Most - about 75%

Classification of damage to buildings

Grade 1   Slight damage: Fine cracks in plaster; fall of small pieces of plaster.
Grade 2   Moderate damage: Small crack in walls; fall of fairly large pieces of plaster; pantiles slip off; cracks in chimneys; parts of chimneys fall down.
Grade 3   Heavy damage: Large and deep cracks in walls; fall of chimneys.
Grade 4   Destruction; Gaps in walls; parts of buildings may collapse; separate parts of the building lose their cohesion; inner walls and fill-in walls of the frame collapse.

Grade 5   Total damage: Total collapse of buildings.

Arrangement of the scale Introductory letters are used to paragraphs throughout the scale as follows: a) Persons and surroundings, b) Structures of all kinds, c) Nature

5. General Procedures for Conducting an Investigation

   Getting Started
   Special Problems of Gaining Access to Lifelines
   Prioritizing Different Lifelines
   The Approach to Investigating a Lifeline
   The Approach to Investigating a Facility
   Interviewing Techniques
   Documenting Lifeline Seismic Response
   Documenting Good Performance
   Emergency Response Plans
   Special Situations
   General Site Evaluation Check List

5. GENERAL PROCEDURES FOR CONDUCTING AN INVESTIGATION

Many characteristics contribute to the success of a lifeline investigator but the key factors are inquisitiveness and persistence. While no investigator will be an expert on all of the systems, facilities and equipment that he or she is expected to investigate, persistent questioning coupled with what is known from past earthquakes and the current situation will ferret out the information that is sought.

GETTING STARTED

The reconnaissance team should establish contact points and a general plan for investigating the earthquake. Investigation teams, transportation and translators should be organized. Ideally a team in the field should include two or three individuals with a translator. Remember that in general the lifeline investigators will be part of the EERI or NRC reconnaissance team. It is important to make sure that some member of the team will have responsibility for finding and collecting or making arrangements for getting copies of strong motion records in the region.

If the local phone system is working try to meet with your local contacts identified before leaving the U.S. They will have been on the scene and will probably have a good idea of where the high damage areas are and important facilities that have been damaged. Try to arrange for a local translator to accompany you. If your contact is an engineer, he will probably be very busy dealing with his clients, but he may be able to provide some leads. University contacts are particularly helpful as a graduate student may be found to assist. Ideally, it would be desirable to get someone with a technical background to reduce the problems with technical translation. A lack of translators may require larger investigation teams than is desirable.

Get recent local newspapers and look for special publications about the earthquake. Very often a book, primarily of damage pictures, will appear a few days after the earthquake. Attempt to get maps of areas that will be visited.

SPECIAL PROBLEMS WITH GAINING ACCESS TO LIFELINES

Frequently, when investigations are organized, EERI or NAS establish local contacts in the country that is to be visited, typically individuals or organizations with a Civil Engineering orientation. Likewise, university interpreters will often have civil engineering backgrounds. While the host organization will often respond to requests to establish contacts with lifeline organizations, it is suggested that an independent effort be made to establish contacts. There would be four approaches to this. First would be through contacts provided by team members. Others include contacts made through the appropriate departments in the local university, through professional organizations, or through direct telephone contacts. Clearly arranging contacts must be done in a way that does not offend the host organization.

For foreign earthquakes, it is desirable for the investigation team to have a member to concentrate on government contacts. This individual would spend most of his time at the center of government, even though that might be far removed from the earthquake. Another alternative is to visit the government center, often the capital, on the way home before departing the country. This individual should observe the impact of the disruption of lifelines on the governmental response. In addition, many lifelines are operated and/or regulated by the government so that government officials will have information on lifeline response and can also assist in gaining access to lifeline facilities.

PRIORITIZING DIFFERENT LIFELINES

Planning will be difficult because of a lack of information so that one must be flexible and take advantage of opportunities as they become available. Because of the diverse character of the lifelines, information about damage becomes available sooner for some systems. For example, damage to power, communication and transportation systems is the first to surface. Much of the damage to water, sewage and fuel systems surfaces later since much of these systems consists of underground facilities and they usually must be excavated to assess damage. Usually, when an investigation team is at a particular site it will investigate all the lifeline facilities that are there. This is why it is important for investigators to be well versed in all systems.

After a large earthquake where extensive damage is to be investigated, time will be the limiting factor. Each earthquake is unique so that "cookbook" guidance will not work. It is important to focus on the four main objectives of investigations: describe damage, identify failure modes, determine factors contributing to the failure, and determine impacts of the failure (to the system, the emergency response, and the community.) There will never be enough time to gather all the information that is waiting to be uncovered. Thus, it is important to know what has the potential for yielding new observations rather than just reconfirming observations from past earthquakes. What makes this task difficult is that damage that has been observed frequently in the past may have different causes and impacts in the current situation. Since failure modes and impacts of damage have been neglected in the past, most damage will have to be investigated, but the focus may be different than in the past.

Teams will often travel in the same general area and stay at the same hotel at night. Lifeline investigators should plan to meet at least an hour to review the days observations and plan for the following day. Attempt to have the specialist of a particular facility be in the first team to visit the site.

THE APPROACH TO INVESTIGATING A LIFELINE

There are several approaches to the investigation of lifeline damage. Individuals from large organizations may have a tendency to play it by the book. That is, headquarters of the utility will be approached to request permission to visit various facilities. Unless you have a contact at the utility or one of your local contacts does, this is not the recommended approach. The problem is that depending on how long it is after the earthquake, people at headquarters will be very busy and to get rid of you they may tell you that you cannot enter any facilities. Thus, it would now be improper to attempt the entry of any of this utilities facilities. Unless you know what the answer is going to be, it is better not to ask the question. At some point, headquarters must be approached so that an overall view of damage and system response can be obtained.

Most of the time they will be cooperative and will arrange visits to other sites of interest.

While the central or main office of a particular lifeline or utility will be able to provide the best information of system response and major aspects of the damage, the best place to get information on the extent and character of damage is at the damaged facility. For information about the transmission and distribution systems for that lifeline (where this is appropriate, ie, power, communication, water systems, etc.) go to the various service centers (maintenance yard, dispatching yard) where service crews are dispatched to make repairs. The dispatcher's log will have details on each repair, and statistics and spatial distribution can also be gathered. Many of the "small" jobs are not reported to the main office but these in their aggregate can have a significant impact. Information on the availability of spare parts, the number of repair crews that are available, problems that repair crews are having, etc. can also be obtained. A large city will have several of these service centers. There may also be different types of service centers for different parts of the system. For example, for power systems, different centers may be used for the distribution system and for the transmission system.

THE APPROACH TO INVESTIGATING A FACILITY

As noted above, many facilities will be approached cold, that is, one drives up to the front gate and asks to see the plant manager. Very often in the first week after an earthquake there is so much confusion that if you approach a facility directly you can gain entry and gather the data that you want. Normally, the local manager would not let anybody in the facility without clearing it with the main office. Because communications for contacting the main office may be difficult and because the main office will be busy, you may be given direct entry. The fact that the investigators have hard hats gives them an air of being "official" and that they are foreign may also ease entry. A week or so after the earthquake when things have settled down, entry into facilities without proper approval may be more difficult.

Upon entering the facility there is usually a meeting with the manager where business cards are exchanged. The purpose of your visit should be explained and how the information will be used. You might indicate that a copy of your observations will be sent for review before they are finalized. In general, it is desirable to provide each facility with a copy of the observations made at the facility with the idea that the material will be checked for accuracy and completeness. This, however, is a time consuming process, particularly for foreign earthquakes. Within the U.S. this should be standard practice. This follow up communication provides an opportunity to ask specific questions that may have been raised when working up the material later. You might enclose a print of the picture that you took of the individual as a courtesy (see next paragraph).

The manager may give an overview of what happened at his facility. It is important to remember that questions about damage and disruption may be embarrassing or threatening. While judgement must be used in asking questions, it may be better to look at what was damaged and reserve your questions on the impact of the damage and the restoration process until after touring the facility. (For some lifelines, such as power and communications systems, the effort to remove damaged items and get things operating again will usually start within an hour after the earthquake.) Also, after reviewing damage at the facility you will be better able to understand and ask questions about various aspects of the response. Before starting the plant tour, it is best not to ask to take pictures. The manager will see your cameras and if he does not want you to take pictures he will have the opportunity to make his position known. It is not uncommon that when a facility is visited a second time by you or another group that pictures will not be allowed even though they were taken on the first visit. It is a good idea to take a picture of each individual who provides significant information while in the plant and to take a picture of the sign at the entrance to the plant. This helps to correlate notes with slides.

At the first meeting with the facility manager he may be much more willing to talk about how disruptions of other lifelines affected his operations. For example, did his facility lose power, water, communications, etc. and for what duration? What was the impact of the disruption on his operations? You might also seek information that he may have about other industrial facilities or other facilities within the utility. Try to set up a second meeting with the manager before you leave. At that time you can ask questions about the impact of various items that were damaged and special things that may have been done to get back into service. Detailed questions about the cause of damage should be avoided as the manager may not know the details and it may be embarrassing to him. At the closing meeting, the manager can be requested to set up a visit to another damaged facility by calling them directly. This will assure that you get in and set the stage for a successful site visit. Try to find out the age of the facility and if any special seismic design specification are used for the purchase of equipment or its installation.

When in the plant, typically, you will be guided by a manager or supervisor. Attempt to ask questions of the individuals who actually did repair work as they will know the details. Also ask questions about problems in other parts of the plant. If you know of a problem associated with the area you are dealing with you might ask a pointed question about it rather than just asking about damage in general. Questions should be asked about signs of damage or repair.

It is important to get an estimate of the character and severity of the ground motion at each site that has significant damage, even though there will be others on the reconnaissance team who will be characterizing the ground motion and attempting to get copies of strong motion records. Since the character of ground motion can be quite variable from location to location and most sites will not have a strong motion recorder near by, it will be necessary for investigators to characterize the motion from observations at and around the site and by interviews with people who were at the site during the earthquake. Within the U.S. the Modified Mercalli Intensity Scale is used to characterize the effects of the earthquake. Appendix A, Field Guide for Lifeline Earthquake Investigations, should be used to characterize the local intensity. If possible and if the ground conditions are similar to those at the site, adjacent areas should be used to help determine the intensity.

Also, since the site seismic environment is strongly influenced by soil conditions, important characteristics of the site should be noted. For example, is the area adjacent to the site flat or hilly, are parts of the site on fill or cuts, is there evidence of soil deformations such as subsidence or liquefaction, etc?

Some questions in interviews with people who were at the site at the time of the earthquake should be directed at characterizing the intensity. Note that the perceptions of individuals are related to the intensity as described in the Modified Mercalli Intensity Scale.

INTERVIEWING TECHNIQUES

In general the use of small teams of two or possibly three (plus interpreter) for conducting interviews is preferable. With smaller teams there can be more teams and more sites can be visited. Also, a larger team can be intimidating to the people at the facility you are visiting. In general, one person will pursue a line of questioning while the other will take notes. While these roles may switch, each team will work out the procedure that works best for them.

It must be emphasized that any questioning may seem threatening and this can be accentuated by the tone and phrasing of the questions. The purpose of your investigation should be explained to each guide if more than one takes you through the facility. Even if there are no language problems, there are often communication problems. Be particularly wary if a term appears to be used in a slightly different way than you are used to. The importance of the "inner ear" to sense misunderstanding or holding back of information cannot be over emphasized. Even if you do not think that there is a misunderstanding, for important issues, ask the same question more than once in a different way at a later time or of different people. Persistence is needed to get things clearly understood. Frequently there is a reluctance of plant personnel to bring up the existence of problems, but, if you can bring a problem up from sketchy information that you have gotten from others, an open discussion often follows. As a source of information for later use, ask about other areas outside of the authority and responsibility of the individual who you are with. There is often less reluctance to talk about other peoples' problems. While one does not want to be patronizing, when you observe good seismic practice mention it and explore what brought it about.

DOCUMENTING LIFELINE SEISMIC RESPONSE

Documenting Damage - Documenting earthquake damage is the first and easiest step of the four areas of investigation that are being pursued. It is important to identify the manufacturer, model number, and equipment description including its rating, capacity, size, age, operating status at the time of the earthquake, use of seismic specifications in its design or procurement, etc. While damage to major facilities and equipment are usually mentioned in interviews, many "small" things are over-looked or are considered inconsequential. It is important to seek these out, since the failure of a minor item may have significant consequences or the aggregate of many such failures may be significant. When there is equipment damage at a site where there are several items of similar equipment, information of the percentage of equipment that is damaged should be gathered.

Failure Modes - Knowing how many items of a particular piece of equipment failed is not enough information to be of use in preventing similar damage in the future. The failure mode should be identified or as may be the case in an earthquake investigation, speculation as to the failure mode should be given. Since the investigator is on the scene, he will have the best opportunity for determining the failure mode. One of the best sources of information will be personnel at the site; however, these people are primarily concerned with the operation of the entire facility and may not know much about the operation and construction of individual equipment items. If possible, try to speak to the people who do the mechanical maintenance of the equipment. More is needed than just identifying the failure mode. For example, indicating that a ceramic column on a high voltage circuit breaker failed does not suggest what factors contributed to the failure.

Factors Contributing to Equipment Failure - What are the causes of damage? Again, it may not be possible to make a definitive statement as to the cause of damage, but you may be able to provide some guesses. Provide as many details as possible. Consider the example of the damaged circuit breaker referred to above. Was there a lack of slack in conductors connecting adjacent equipment so that relative displacement overloaded the ceramic member? Did the anchorage fail or contribute to the failure? If there was an anchorage failure, what are the details of the failure? Was the item anchored, did a weld fail, did anchor bolts shear or pull out, was there damage to the concrete around the anchor bolts, what was bolt diameter, length and type (cast-in-place or inserts), was there excessive deformation of the equipment near the anchor points, etc? Even if the anchorage did not fail, excessive flexibility could have contributed to the failure. It is only with this type of detailed information that informed engineering judgements can be made as to the cause of failure. Frequently the damaged equipment may have been replaced by the time the investigation takes place. If the equipment has been removed, can you get pictures of nearby, similar equipment that did not fail? Damage is usually documented with pictures. The above questions also serve as a guide for the type of pictures that are required. If a site contains similar equipment in which some is damaged, differenced is design, constructions, installation or location on the site should be investigated in an attempt to determine the differences in the response.

Attempt to find out the age and date of installation of individual items that are damaged. This may be on name plates of large equipment items. Look for good and poor seismic design practices.

Because of the limited time to investigate each failure, notes and pictures made during the investigation often must be used to piece together the entire story of the earthquake response after field work is completed. Pictures which carefully document the damage scene are indispensable to this process. Thus, detailed pictures of the damaged equipment including close-ups of fractured and deformed parts, connections to adjacent equipment and anchorage details must be collected as well as overall views from different perspectives. It is here that a zoom lens can make a big difference in the quality of the damage pictures. A macro lens can also be useful when visiting a site at which pictures were taken immediately after the earthquake by facility personnel. Requests for prints of these pictures are usually fruitless however the use of a macro lens can yield good results.

What Are the Impacts of Damage - Equipment damage can impact an entire facility which can, in turn, disrupt the entire lifeline. Lifeline system malfunction can disrupt other lifelines, all of which can be very disruptive to the emergency response and the community at large.

What were the effects of equipment failures on the operation or performance of the facility? Some damage can be quickly repaired or the equipment can be bypassed so that its loss does not cause much disruption. Other failures may have major impacts. Determine the impacts of damage and if the particular situation or configuration played a role. In some situations the facility can continue to operate but it must operate in a different way. What was the impact of facility impairment on the system operations and performance?

What were the secondary effects from lifeline damage? What were the effects of disruption beyond the system and what measures were used to circumvent problems? For example, loss of power may shut down water pumping stations with a large drop in water pressure. Do pump stations have emergency power that worked, does the fire department have pumper engines and/or stand alone pumps to draw water from local sources such as a river, lake or ocean? How have critical facilities and services operated with the loss of lifeline services? Do emergency response centers have back up communications, do hospitals have back up power and water, etc.

DOCUMENTING GOOD PERFORMANCE

The primary focus of earthquake investigations is to document damage and its impact on systems and the community at large. However, good performance is also important so that when a site is known to have experienced significant ground motions, equipment which has performed well should be documented. Just like damaged equipment, the manufacturer, model number, and equipment description including its rating, capacity, size, age, etc. as well as pictures should be recorded. With the new policy of getting several people into the field to gather data, more information on equipment that has performed well should be collected. Two types of information on good performance are useful. Find out the number of similar or identical equipment items at a site are damaged and that are undamaged. If you gather detailed information about equipment that was undamaged, it is important to have a good idea of the level of ground motion at the site. It is also important to compare similar equipment items that were and were not damaged to see if differences might explain the failures.

EMERGENCY RESPONSE PLANS

Inquire if the organization and the facility have emergency response plans and if there are any special provision for earthquakes. Ask if the plans are written and if they are ever exercised or evaluated.

SPECIAL SITUATIONS

It may turn out that the Lifeline Investigations Committee will initiate an investigation independently of EERI and NRC because they decide that the earthquake does not justify an investigation. This can be the case for small earthquakes in which damage is very limited. For example, in the recent Tejon Ranch earthquake, damage was primarily limited to a switchyard associated with the California Aqueduct and to the aqueduct.

If the Earthquake Investigations Committee mounts its own investigation, then it is important to pay special attention to characterizing the intensity of the earthquake. This type of investigation will probably be limited to the U.S. and organizations such as the U.S. Geological Survey will investigate, even small earthquakes.

GENERAL SITE EVALUATION CHECK LIST

- What is the orientation of the site relative to magnetic north?
- What is the topography around the site?
- What parts of the site are on cut or fill?
- Can an estimate of the depth of the cut or fill by looking at the boundary of the site?
- Is the site on alluvium? Inquire as to its depth.
- What are soil conditions at the site: rock, very firm soil, firm soil, soft soil?
- Cornell Soil description
- What is drainage of site, above or below grade?
- What is the depth of the water table?
- Is there evidence of soil deformations?
- Does the site manager know if there was any special foundation preparations at the site?
- Did you get an estimate of the MM Intensity for the site.
- Did you get a card for the plant manager with his address?
- Did you evaluate the over all quality of construction and the use of good seismic practices?
- Did you estimate the percentage of failures for each type of equipment.?
- What was the extent of disruption at this location?
- What was the duration of the disruption at this location?
- What was the time to restore service and to complete repairs?
- What was the impact of dysfunction at this site on system?
- What was the impact of dysfunction on other lifelines?
- What was the impact of dysfunction of the lifeline on the community?

6. Electric Power Systems

   System Configuration
   System Operations
   System Facilities:
      Their Function , Seismic Performance, and
      Earthquake Investigation
   Facilities and Equipment with Good Earthquake
      Performance
   Case Study -
      Investigation of Tejon Ranch Earthquake/
      Edmonston Switchyard
   Power System Check Lists
      Anchorage Check List
      Power Equipment Check List
      Power Plant Check List
      Substation Check List

6. ELECTRIC POWER SYSTEMS

SYSTEM CONFIGURATION

From an overall physical perspective, power systems consist of a number of nodes (substations and power plants) which are typically interconnected by redundant networks of transmission and sub-transmission lines forming a grid network, often called loop systems. Emanating from some nodes (distribution substations) are radial systems (tree networks) of feeder lines and service lines that carry power to users. It should be noted that nomenclature used to define some parts of the system varies between utilities within the U.S. and from country to country. It should also be noted that in the descriptions that follow, typical facilities and situations will be discussed. Some utilities organize their systems in different ways to better address their particular needs. The objective of the following descriptions is to provide the earthquake investigator with some familiarity with power systems and not to create a power system specialist.

Before describing the elements that make up a power system, it is important to understand a few characteristics of power networks that distinguish them from other lifeline networks such as transportation or water system networks. First, power systems generate and distribute a commodity, electric power. The system does not store power (in its electrical form) so that it is used as soon as it is generated. Second, there is little control over the flow of power within the network once it has been generated. Electric power generated at a power plant will travel along all of the lines which interconnect the generator and the user. The current flowing over any path is inversely proportional to the impedance of the path; the lower the impedance the higher the current. The capacity of a path is not necessarily related to its impedance, particularly if the system is configured in an unusual way. If the current along any path exceeds the capacity of any element within the path, protective monitoring devices will cause a circuit breaker to open. This will stop all current in the path and alternative paths must carry the current formerly carried on the opened circuit. To balance power generation and consumption and to control the flow of current through the system power output of the generating stations distributed throughout the system can be adjusted, the configuration of the transmission network can be changed, and in extreme cases load can be dropped. The inability to store electrical energy in the system and the immediate reallocation of the flow of energy within the system to any changes in the system configuration requires a sophisticated and sensitive control system to provide reliable service in the face of numerous problems that commonly befall power systems.

For the purpose of earthquake investigations, power systems can be divided into five major parts: power generating facilities, transmission and distribution lines, transmission and distribution substations, control and data acquisition systems, and ancillary facilities and functions. Within the U.S., the country is divided into 7 reliability councils. Within the Western US the Western States Coordinating Council (WSCC) has been formed to address regional system problems. The region includes all or parts of 11 states and extends into Canada and Mexico. A map of major transmission lines of the WSCC is shown in Figure 6.1. This clearly shows the extensive and redundant character of the power transmission system. The large concentration of lines in major metropolitan areas cannot be depicted at this scale. Details for Los Angeles and San Francisco are shown in Figs. 6.2 and 6.3. While most of the power grid within the U.S. is strongly tied together, the WSCC is unique

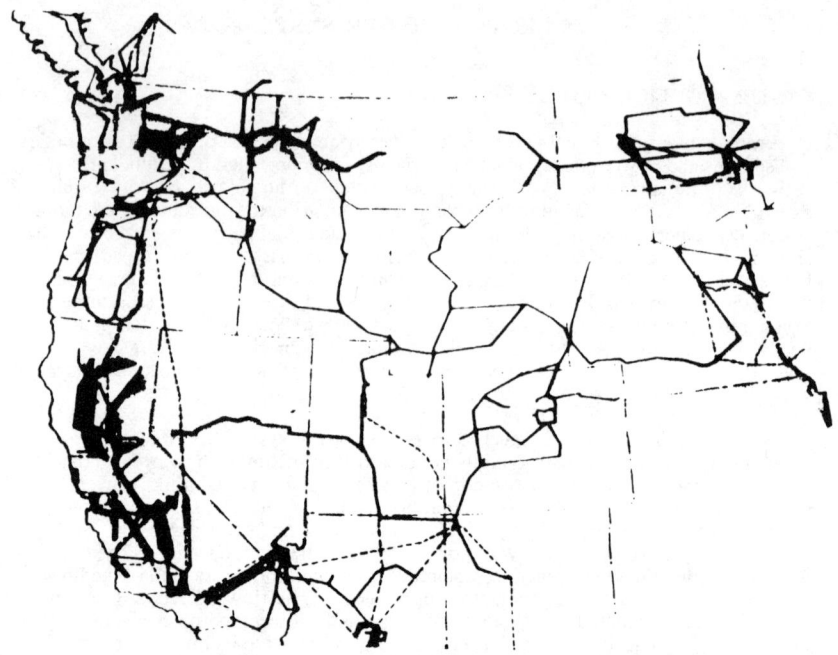

Fig. 6.1 Power network in the Western States Coordinating Council

in that it operates nearly independent of the rest of the country. In strongly tied networks line frequency (60 Hz in America and half of Japan and 50 Hz in most of the rest of the world) is synchronous; that is, it operates in phase.

Figure 6.4 shows a schematic diagram of a portion of a simplified power network. The power output of a generating station passes through a step-up transformers located adjacent to the generating unit. The function of this transformer is to increase the voltage from the generating voltage (2 kV to 24 kV) to transmission voltage. A substation near the generating plant serves to connect the plant to the power transmission network. Power from this and other generating stations now flows through the system. Note that in the simplified system shown in Figure 6.4 there are two transmission voltages, 500 kV and 230 kV. Transmission substations serve to connect several transmission lines, all at the same voltage or to transfer power between different transmission voltages levels within the system. The thickness of the lines in the figure is an indication of the voltage at which each operates. Distribution substations are connected to the transmission system. Very often a large utility will have another network at distribution voltages. Some utilities will refer to the higher voltage network as their transmission system, the intermediate voltage system as their sub-transmission system. Note that at a single location a facility may have different parts that serve as transmission, sub-transmission and distribution substations. The power system shown in the figure would support many more distribution substations but they were not shown for simplicity.

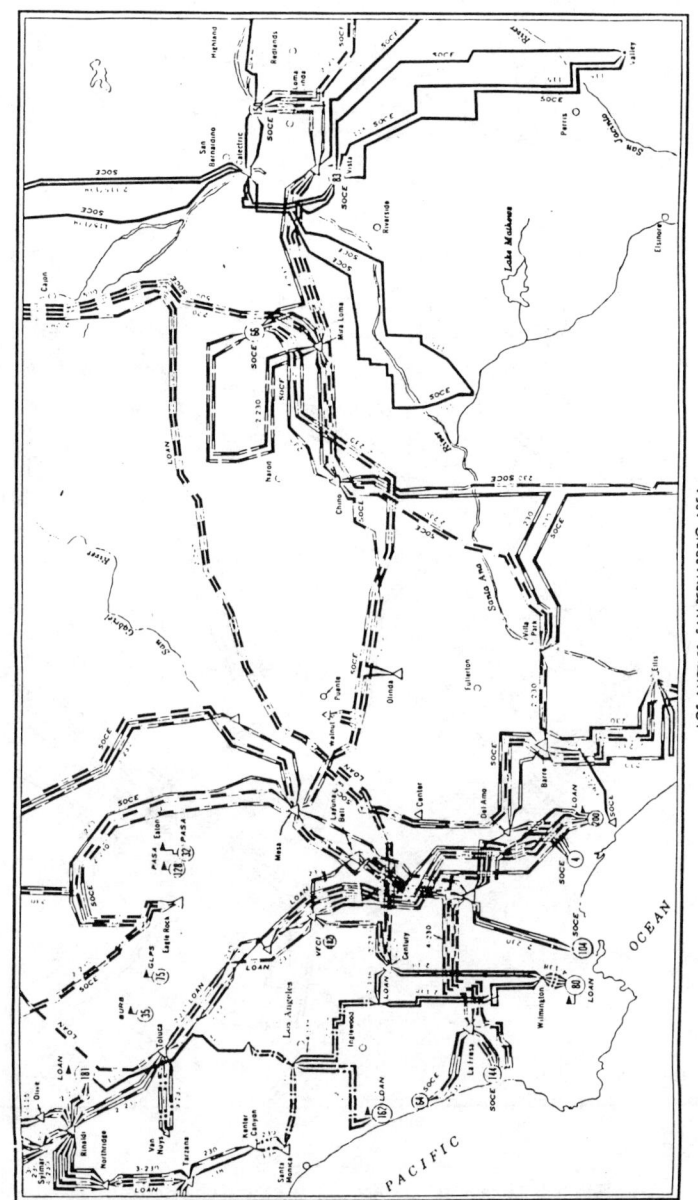

Fig. 6.2 Transmission system in the Los Angeles area

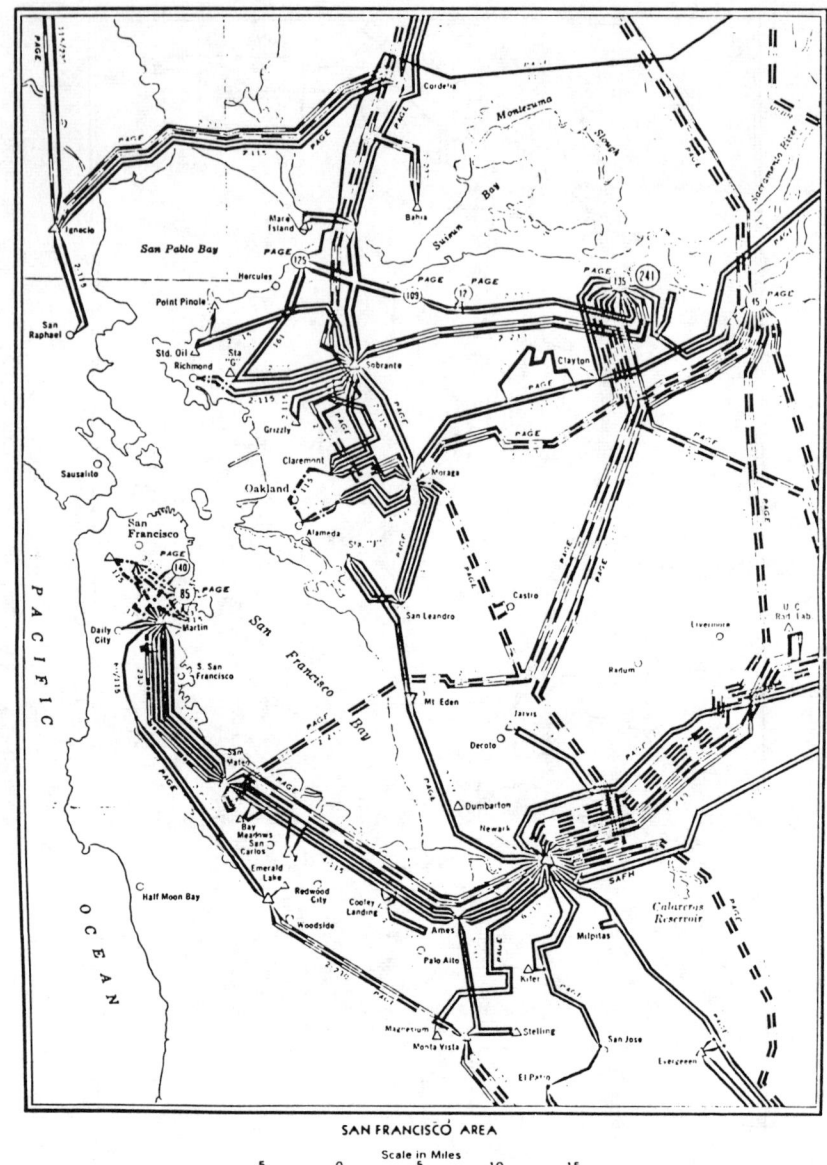

Fig. 6.3 Transmission system in the San Francisco area

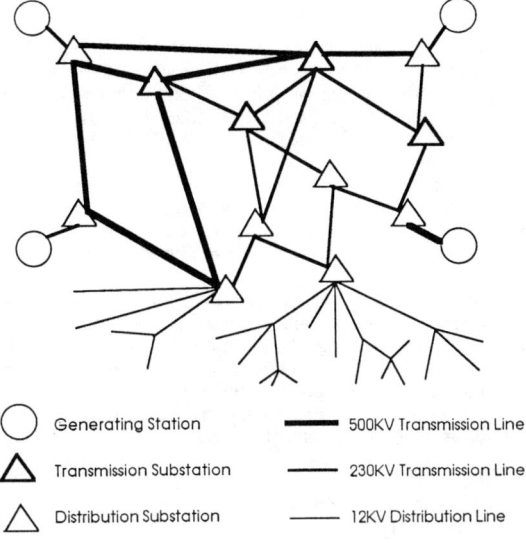

Fig. 6.4 Simplified schematic diagram of a power system

Distribution substations will convert power from the transmission or sub-transmission levels to distribution voltages. Besides connecting distribution lines, the output from most distribution substations goes into feeders that carry power through a radial network to the users. The feeders may provide power to vault transformers at intermediate size customers such as shopping centers, industrial facilities, and large buildings, where it may be converted into one of several possible voltages, usually 2.4 kV, 480 V or 220 V. Pole- or platform-mounted transformers typically supply service connections to individual residences at 240/120 volts.

Each utility typically has several transmission voltages within its system. In general, higher voltage lines carry more power, cover longer distances and experience lower losses. Transmission voltages range from 110 kV to 800 kV. Within the U.S. the highest voltage is 765 kV and it is limited to the Midwest. More widely used is 500 kV. Several other transmission voltages are used, the most common are 345 kV, 230 kV and 110 kV. Some utilities would consider 110 kV as a sub-transmission voltage. Sub-transmission voltages can range from 60 kV to 110 kV. The voltage on distribution feeder lines ranges from 4 kV to 44 kV, the most common feeder voltages are in the range from 4 kV and 21 kV. There is a tendency for voltages in a system to increase as facilities are upgraded to carry more power over existing right-of-ways. It should be noted that the voltages referred to are nominal values and actual values may be different.

SYSTEM OPERATIONS

The center of power system operations is the control center. Two key functions are directed from here: power dispatch and system configuration. Because total power generation must match consumption, the output of generating units must be continually adjusted. From an economic point of view it is desirable to have the least efficient generating units supply the marginal power demand so that they can be throttled back as demand falls. However, units must be distributed so that the flow in the system is within the capability of transmission lines. The system is monitored and generation is adjusted on an ongoing basis, sometimes as often as every 10 seconds, through computer dispatch. An integral part of the control system is the communications system required to monitor the status of the system and to send commands to generating units.

The dispatcher at the control center also makes decisions about changes in the configuration of the system. Lines are opened for service or system needs or a segmented bus (to be discussed later) may be partitioned. The actual opening of lines may be done remotely or manually by personnel in the field. Individual lines may open through the action of a circuit breaker upon sensing a fault, such as a short circuit to ground, for short periods until the fault is cleared. The decision as to what generating units should be operated is done in the control center.

SYSTEM FACILITIES: Their Function, Seismic Performance and Earthquake Investigation

Overview

Before describing individual facilities and their seismic performance, overall seismic performance of power systems will be reviewed. In addition, common failures and factors which contribute to them will be discussed.

Numerous recent earthquakes have demonstrated that certain elements of power systems are extremely vulnerable to earthquake damage while other elements are highly resistant to seismic damage. These notes will focus on those elements that have been shown to have poor performance or which are suspected of being vulnerable. Good performers will be noted but not discussed.

In many recent moderate earthquakes there has been significant damage to ceramic members of high voltage (≥ 220 kV) substation equipment. Since the earthquakes were of moderate magnitude, seismic induced shaking was limited to a relatively small area and significant damage was usually limited to a single substation. Because of redundancy incorporated into power systems, system performance has been good even though a substation was severely damaged. Existing damage patterns clearly indicate that a larger earthquake, or a similar earthquake in regions of the country with lower attenuation of seismic waves will impact several stations and the damage will likely overwhelm the ability of the system to respond. Since the evolution of modern power systems, no large or great earthquakes have struck a modern metropolitan area, so that power systems have not been subjected to long-duration, high intensity shaking. Thus, the performance of individual equipment items, the facilities of which they are a part and the power systems as a whole to these conditions is not known. The earthquakes that caused power system damage to date have had durations of less than fifteen seconds and many lasted only a couple of seconds. While ground acceleration in an epicentral region can exceed 1/2g, significant high voltage substation equipment damage has occurred at accelerations as low as 0.05g.

Common Failures

Five types of failures dominate power system damage. The most frequently observed and most serious is the failure of ceramic members in high voltage substation equipment. Other common failures include equipment anchorage, battery racks for station batteries, various types of liquid storage tanks, and distribution lines. Since liquid storage tanks are common to many lifelines, they will be discussed separately in Section 13, Common Facilities. Distribution lines are discussed later.

Failures of Ceramic Members

Because of the desirable electrical properties of ceramic materials, they are found in virtually all high voltage power system applications such as bushings, post insulators, suspension insulators, lightning arresters, circuit breakers, potential transformers and current transformers. Unfortunately, the mechanical properties of ceramic create severe problems when it is subjected to earthquake induced loads.

The tensile strength of ceramic is much lower than materials such as aluminum and steel which are typically used as structural members. The main problem, however, is the brittle character of ceramics, which has two serious effects on the structural performance of the material. In most common structural materials, which are ductile, a localized stress will cause local yielding so that the stress is redistributed over a larger volume, thus local peak stresses tend to be limited. This does not happen in ceramic. In ceramic materials, when the stress at any point within the members exceeds a threshold level, the member will fail catastrophically. Another factor that complicates the use of ceramics and the analysis of failures is that in the manufacture of ceramics, voids and inclusions are incorporated into the body of the material. These act as local stress raisers so that a member may fail when the overall stress is relatively low. The uncertainty of the number, size and distribution of theses flaws means that ceramic members exhibit large variations in strength as compared to traditional constructions materials, such as steel and aluminum

Unfortunately, ceramic members are usually configured as vertical columns with electrical connections or masses supported at their upper ends. This has the potential of placing large transverse loads at the free end which translates into large moments at the base of the column. These columns are assembled in one of two ways. The most common type of support has a metal flange bonded to the base (bushing, or lightning arrester) or ends (post insulator) of the ceramic member. In this case the column becomes a cantilever beam and the large moment at the lower end of the beam translates into stresses in the outer fibers at the base of the member while most of the material throughout the rest of the member experiences substantially less stress. The other type of assembly has internal members that are in tension (tendons) that clamp the ceramic members together (some types of bushings and support columns on some live tank circuit breakers). The ceramic members, which are often pressurized with an insulating gas, are seated with a rubber like "O" ring. In this case the large moment at the base of the column may reduce the compressive load on the "O" ring and allow the internal pressure to blow out the ring from its seating. This will prevent the device from operating as it was designed and will allow ceramic-to-ceramic or ceramic-to-metal contact that can create a stress concentration and ceramic failure.

In an earthquake investigation it is desirable not only to document the number of failures, but also to determine the most likely cause and factors that contributed to each failure. This will typically require knowledge of the seismic loads induced in the member. Three basic loadings are possible: inertial, imposed relative deflections, and impact induced loads. Each has several factors which influence it.

A piece of equipment that moves with seismically induced ground motions will experience inertial forces. The basic ground accelerations can be amplified by the dynamic response of the equipment, by the dynamic response of the equipment support, and by the dynamic response of the structure in which the equipment is mounted. The inertial forces are distributed over the equipment and will be proportional to the mass and acceleration. Member loading can be estimated by use of equivalent static analysis where peak ground accelerations are scaled to account for the system amplifications. Damping will be a function of the equipment and can range from 2% to 5% of critical. It should be noted that for short duration excitations, the system will not have time to reach its steady state response.

A second source of loads and probably a source which is more severe than inertial loads are those that arise from the interaction between moving equipment items or parts of the same equipment item. Substation equipment is usually interconnected by electrical power conductors, usually referred to as bus. To carry the required power these conductors are often an inch or more in diameter so that they can transmit large forces and tend to be relatively rigid unless special provisions are made to add flexibility. Consider a current transformer and an adjacent circuit breaker. Each piece of equipment is mounted to a support structure that may be seven feet high and the equipment may be twenty feet tall. Flexibility in the support structures and the equipment, particularly if the equipment can rock, will translate into large deflections at the top of the equipment where the bus is usually connected. If the slack in the electrical connection between the equipment cannot accommodate the combined peak deflections of both pieces of equipment the bus will pull on the equipment, usually a tall ceramic member.

In addition to the above source of interaction loads, there are other potential sources. There can be interaction within a single piece of equipment. For example, multi-head, live-tanks circuit breakers may have interaction between heads supported on different columns. Another example would be loads applied by strain bus due to the dynamic response of the bus. Strain bus, or flexible bus is fabricated from twisted aluminum conductor rather than from aluminum pipe, which would be referred to as rigid bus. Also, when equipment moves due to inadequate anchorage, for example, the sliding of a transformer, connections to the equipment may load and damage bushings.

Nonlinear and impact induced loads can also influence the loads that ceramic members will experience. Even minor impacting can create very large loads and the source of the the impacting can be quite varied and often difficult to identify. Some of those that have been observed associated with anchorage include lack of shims under an anchor point, damaged or missing grout under an anchor point, and lose anchor bolts. When any of these conditions exist, impacting at the anchor points can occur at relatively small earthquake motions. Another source of impacting is in equipment support structures which use bolted connections with slots rather than holes. Under seismic loads, even tight bolts will allow bolted members to slide and when the bolt reaches the end of the slot a severe impact is imparted to the entire structure. Connections with inadequate slack tend to impart an impact when the slack member becomes taut. A similar effect occurs when an object supported by an insulator string or with guys experiences an earthquake. Tendon or spring loaded ceramic members tend to impact when seismic loads exceed pre-loaded assembly loads. This can be very severe when gaskets or "O" rings are blown out so that a ceramic member will come in direct contact with another ceramic member or metal gasket seat. In addition to creating large loads, impacting and nonlinear loads can excite higher modes of vibration of the equipment.

Failures of Equipment Anchorage

Much of the observed power system damage can be attributed to the lack of or failure of equipment anchorage. Inadequate anchorage allows equipment to move or fall over. For equipment which moves, a lack of flexibility or slack of connections to nearby equipment can cause the connections to break or load the equipment so that it fails. Examples would be failed piping, broken bus connections, fractured pump nozzles, and fractured bushings. Equipment which falls over is typically damaged by the fall rather than from earthquake vibrations that it experienced. The failed anchorage, per se, usually does not cause the equipment to malfunction.

It should be emphasized that anchorage as used here refers to the entire load path from the equipment to its support and must take into account the stiffness of the support structure and the equipment near anchorage points. As noted earlier, a flexible anchorage may cause rocking of the equipment that can include large deflections at bus connections near the top of the equipment. Another problem with flexible anchorage is that it may lower the natural frequency of the equipment so that it is more attuned with the high energy frequency content of the earthquake (0.5 Hz to 6 Hz). It should be emphasized that detailed analyses of equipment damage frequently is tied to the flexibility of the equipment and its support structure.

Some of the questions about anchorage include the following. Are the anchor bolts cast in place or expansion anchors? What is their length of embedment? What is their diameter? How did they fail? Did they pull out of concrete? Did fracture cones develop in the concrete? Did the bolts stretch? Did the bolts break? Is there any indication that they were installed incorrectly? What were the standards, if they existed, when the equipment was installed? How many bolts were there and how were they laid out? Did the bolt pass through a structural member in the equipment framing? Are there signs of distress in the equipment in the region around the anchor bolt? Does the equipment introduce a prying action to the bolt? Is the bolt hole appropriate to the bolt diameter? Does the load path from the cabinet frame to the bolt allow flexibility in the anchorage system? What are the sources of loading on the anchorage: equipment weight, height of center of gravity, dimensions of the base of the cabinet; were there loads applied through interconnections to adjacent equipment? Many of these questions can be answered by a quick on-site inspection. Information about others may be obtained from the design office in the company headquarters. Many questions will remain unanswered due to the lack of time, the inability to access operating equipment, or the inability to see critical parts. In conducting an investigation it is desirable to make informed judgments about the situation and to note the character of your estimate. As an investigator, you may be the only one at the site who will address the issue of what happened and the information that you provide may be all that is available to evaluate the damage. The list of questions also suggest the type of documentation that should be gathered. Thus, pictures of the damaged bolts and of the cabinets in the bolt area will allow the data to be revisited later.

Damaged anchorage may allow the equipment to impact which can contribute to other problems. As noted above, impacting can cause high peak loads that can damage ceramic members. Impact loads may also cause relays within the equipment to chatter; that may cause undesired equipment actions. Review Anchorage Check List.

Power Generating Stations

Power generating facilities can be grouped into four classes of systems. Nuclear-fueled steam generating units, gas turbine peaking units, fossil-fueled steam generating units, and hydroelectric units.

While nuclear facilities generate about 15% of the power in the U.S., because it is very unlikely that access to these facilities can be gained, they will not be discussed further.

Gas turbine generating units are inherently rugged and no seismic damage has been reported. However, the extent to which this class of power generating facilities have been exposed to damaging earthquake is not known. While the turbine/generator is rugged, ancillary systems and equipment such as fuel tanks and control panels could be damaged and cause lengthy disruptions.

A description of the operation and components of fossil-fuel steam generating stations will not be attempted here because their earthquake response has been good to date. Observed damage and suspected damage has not related to the system operation or configuration, except as noted below. It should be noted that large coal-fired plants have not been subjected to damaging earthquakes and even gas- and oil-fired units have only seen limited exposure. Most of this has been relatively small, less than 40 MW (40,000 kW). Larger units have been exposed in Japan, however, seismic design practices are generally used there and are quite different than in the U.S. For example, boiler support structures are much more rigid than in the U.S. and horizontal design loads are typically higher.

While all types of fossil-fueled steam generating units have many things in common, from a seismic point of view fossil-fueled steam generating units can be divided into two categories: oil- and gas- fired units and coal-fired units. Because of the combustion process, a coal-fired steam-generator is larger than an equivalent gas- or oil-fired steam-generator. Also, coal-fired units have massive coal storage silos high in the boiler structure which may increase their seismic vulnerability. They also have extensive coal handling equipment which is often designed without consideration of seismic loads and which has several features which make this equipment vulnerable to earthquake damage.

Seismic Performance of Power Generating Stations and Their Evaluation

The seismic performance outlined below is to inform you of damage that has occurred in the past and to give you guidance as to what to look for in the field. Damage that did not cause a failure can be important and it is indicative of incipient failure. Also, while a particular type of damage may have been observed before, the causes of the damage and its impact may not have been recorded. Thus, it is important to at least briefly review all damage to see if better information on cause and impact can be gathered.

All types of thermal power plants require several large liquid storage tanks and these have proved to be vulnerable to earthquake damage. Since tanks are found in many lifelines, they will be discussed in a separate section. The next most common failure associated with power plants is the anchorage of various types of equipment cabinets. Cabinets with narrow bases are the most vulnerable as their configuration generates large overturning moments and uplift on anchor bolts. Anchorage considerations were discussed above.

Emergency Batteries

Within a power plant the emergency batteries have several critical functions. Should there be a loss of on-site power (a common occurrence in earthquakes due to the action of sudden-pressure circuit relays in transformers or damage to transmission system equipment) the emergency batteries will keep critical communication and plant instrumentation operating and provide the power to start emergency generators. Batteries often also provide the power to operate the lubrication pumps for the turbine bearings when the turbine slows down. The batteries also power the turbine turning gear during cool down to prevent warpage to the main turbine rotor. Often battery racks are not capable of withstanding the large seismically induced lateral loads. Frequently batteries are not secured to their racks so that they can fall off of the rack or get bounced around within loose fitting restraints. Batteries should be restrained from front to back and longitudinally and there should be spacers between cells so that bus connections are not required to restrain the lateral loads experienced by the batteries. Adequate slack should exist to connections between racks and other equipment. Battery chargers, inverters and racks should also be anchored. Damage to battery cases and terminals should be checked. If batteries are lost, it is important to investigate other consequences of the loss of power.

Turbine Pedestal-Powerhouse Interaction

The turbine is typically supported on a massive, stiff turbine pedestal which has its own foundation. At the the turbine floor level, which will be several floors above its foundation, there is usually a construction joint that separates the powerhouse turbine floor from the turbine pedestal. The powerhouse structure is usually much more flexible than the turbine pedestal so that if the construction joint is not large enough, there can be impacting between the pedestal and the powerhouse. While local spalling at contact points may be observed, the real damage from impacting can be to the thrust bearings of the turbine. Also, if there is sufficient longitudinal motion of the rotor, fixed and moving blades can come in contact causing major damage to the turbine.

Turbine Thrust Bearings

Even without the impacting noted above, there can be thrust bearing damage. If this type of damage is observed, it is very important to try to characterize the damaging ground motion so that turbine damage threshold levels can be determined. Probably the critical ground motion characteristic will be deflections along the axis of the turbine rather than peak acceleration or motions perpendicular to the turbine shaft.

Powerhouse Structure and Overhead Crane

Typically one wall of the powerhouse is shared with the boiler structure and the opposite wall is free standing. The typical high bay of the power house means that the free standing wall will be relatively flexible perpendicular to its length and may be damaged. Damage to the overhead crane should be checked as the crane may have acted as a restraint to the flexible powerhouse wall.

Boiler Support Structure-Boiler Interaction

Most boilers are suspended from above and are designed to accommodate large vertical deflections associated with thermal expansion as the boiler heats up and cools down. Lateral restraints may be inadequate. The boiler is often surrounded by truss structure at various levels that serves to resist internal pressures within the boiler. These trusses, often referred to as buckstays, may also serve as a means for providing lateral restraint to the boiler due to close tolerances between the buckstays and the boiler support structure. The restraints may be part of the design or framing members of the boiler support structure may unintentionally act as boiler restraints. On open boiler structures, ties are provided between the boiler (frequently anchored to the buckstay) and the boiler support structure to resist wind loads. These ties are often inadequate for seismic loads and may fail and serve as battering rams to penetrate the boiler. Damage to various connections and interference between members that can not accommodate the lateral motion of the boiler should be looked for. Examples would be highly restrained piping (long run of pipe will be flexible enough to accommodate relative motions) and duct work.

Steam Generator Internal Damage

In 1978, in a large Japanese plant, an internal bracket in a steam generator failed and then served as a battering ram to damage the pipes that it was meant to restrain. This type of damage is indicated by excessive steam in the stack output. While repairs were simple, it took almost a week for the boiler to cool down so that personnel could get access to the damaged area.

Piping.

In general welded steel pipe is very rugged. Indeed, long runs have been observed to experience large displacements and strike structural members with no damage except to the thermal insulation. Some damage has occurred to small pipes with inadequate flexibility which are attached to larger, long pipe run that are more flexible. Threaded pipe connections have frequently failed. Damage has been observed where pipe runs penetrate walls separating different structures so that relative motion between the structures serves as a guillotine for the pipe. Pipes that are anchored at two points that experience relative deflections are frequently damaged.

Systems and Facilities Suspected of Being Vulnerable to Damage

There are some power plant systems in which damage has not been observed, but in which cursory inspections suggest vulnerability. Coal handling equipment, particularly the long conveyor belt support structures that carry the coal from grade or subgrade level to a location high in the boiler support structure would appear to be very vulnerable. These systems are not usually designed to accommodate large seismically induced deflections in the boiler support structure. Another problem may be differential settlement of the turbine foundation. Here differential settlements of a few hundreds of an inch would cause turbine alignment problems.

Substations

Because most substations have similar types of equipment that serve similar functions they will be grouped together for the purpose of earthquake investigations. Substations serve several key functions. Substations provide protection to transmission lines and the equipment within the substation with protective devices for abnormal system operating conditions. Most substations provide for the transfer of power between different voltage levels through the use of power transformers. Substations also provide a means of reconfiguring the power network by opening transmission lines and partitioning multi-section busses. There are other functions performed at substations that support the above primary operations or meet other power system needs. For example, equipment to meet these needs include control consoles, emergency or backup power, and communications equipment.

Power network protective systems are very sophisticated. Protective relays will monitor for over voltage, over current, differential currents, low system frequency, or other parameters. Upon sensing abnormal conditions, operators may be notified or actions are automatically initiated, such as opening a circuit breaker on an overloaded line, opening circuit breakers to isolate a transformer with an internal fault, or opening circuit breakers on feeder lines for over load conditions or for low system frequency. Protective systems incorporate communications and timing circuits so that a given malfunction can be detected and isolated to keep disruption to a minimum. Integral parts of the protective system are communication systems that consist of equipment to communicate over transmission line, utility owned microwave systems and over leased and public telephone system lines. To communicate over power lines, typically a 20 kHz carrier signal is superimposed on one or more power line phases. Line or wave traps (large inductors) used in conjunction with carrier systems are placed in transmission lines within the substation. Certain devices provide passive protection. For example, if the voltage experienced by a lightning arrester exceeds its rating, it will automatically prevent the voltage from exceeding its rating.

Other protective devices that are frequently activated by earthquakes are sudden pressure relays and Buchholz relays in transformers. Large transformers will have one of two types of devices to help sense an internal fault in the transformer and signal circuit breakers to isolate the transformer to minimize internal damage to the transformer. One device is a sudden pressure relay that is a relay in the transformer that senses rate of change of pressure. A sudden change in pressure would be indicative of an internal fault that was arcing, creating gases and increasing the pressure within the the transformer. A Buchholz relay senses rapid flows in oil between the transformer and a storage tank above the transformer. Earthquake imposed vibrations also cause a sudden pressure change or vibration within the relay which is interpreted as an internal fault and the transformer is isolated. This can de-energize parts of the system and cause wide spread outages. If the shaking is mild, the relays may be reset remotely. If the shaking is more severe, the switchyard will be quickly inspected for earthquake damage and, if none is found, service will be restored in about 1/2 hour for a maned site. For an unmanned site, a crew would have to be dispatched to the site to conduct an inspection and to reset the relay if it can not be done remotely. Lightning arresters also provide protection from over voltage in lines by connecting a line to ground if the voltage on the line it is protecting exceeds a given value. When the over voltage condition is removed, the lightning arresters automatically returns to its high resistance state and normal service is restored.

Transformers are the key to transferring power between different voltage levels. Unlike some substation elements, such as a lightning arrester, that can be by-passed and the system can continue to operate without that element, transformers perform a vital function that can not be by-passed. The loss of a transformer will radically effect the operation of a substation unless there is a redundant bank of transformers, or there is a spare on site or a mobile transformer can replace the damaged unit. In recent years there has been a tendency to incorporate the transformer needed for each phase into a single unit, a three phase transformer. In addition to transforming power between two transmission voltages, some transformers also have tertiary windings that may provide station power. It is also very common in the U.S. to have a lightning arrester for each phase mounted on the transformer. Typically, a substation will have some redundancy in transformers. This may be done by having more than one bank of transformers, or , if single phase transformers are used, to have a spare single phase transformer available on site.

Reconfiguring the power network may be required to repair a transmission line or to change the power flow pattern within the network to meet system needs. It is usually done by opening air disconnect switches. These switches however, are not generally designed to interrupt current carrying circuits. Typically, power lines are opened by the action of a circuit breaker. The air disconnect switch would then be opened to deactivate a power line or isolate a piece of equipment. In some substations, a large bus will be divided by air disconnect switches so that the bus can be partitioned. This provides flexibility in configuring the network which may be necessary if the system is severely damaged by an earthquake.

Seismic Performance of Substations and Their Evaluation

Before describing the seismic performance of the elements within a substation, it would be useful to get an overview of a transmission and distribution substation. Two types of diagrams are typically used to describe a substation. The operating single line diagram gives is a simplified schematic circuit diagram of the substation. It gives an overall view of what equipment is at the substation and how it is connected. It does not show the physical layout and it is simplified in that the three phases are shown as a single line. Figure 6.5 shows a operating diagram of a substation that has the elements of a transmission and distribution substation. Table 6.1 shows common symbols used for equipment on operating diagrams. There will be variations in these symbols from utility to utility. This Table also serves as a list of the types of equipment that may be found at substations. Table 6.2 gives translations for common power system terms.

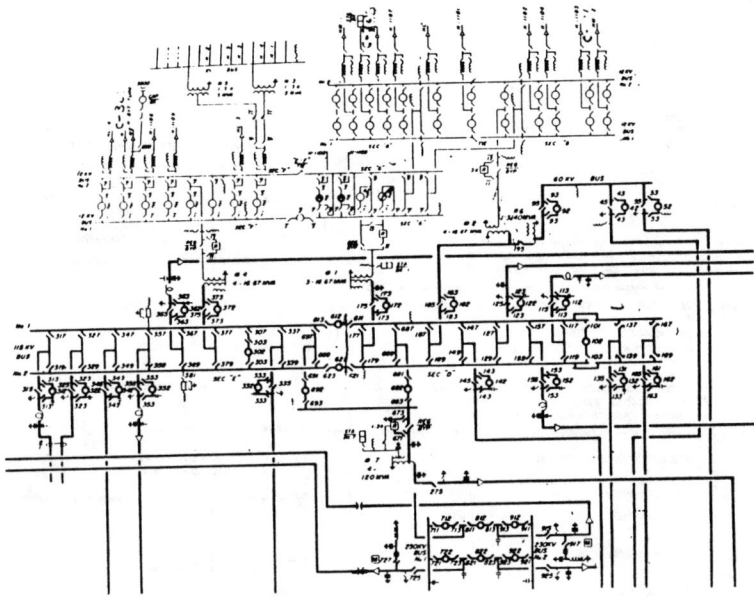

Fig. 6.5 Operating diagram of a substation

A station layout drawing is a second type of diagram of the substation and is shown in Figure 6.6. This type of figure is more useful when looking at a substation during an earthquake investigation. The circuit for each phase is shown and the physical location of all equipment in the diagram corresponds to its located at the site. If at all possible attempt to get a copy of a station layout drawing before conducting the site visit. It is useful to have each damaged piece of equipment circled before entering the switchyard.

It is suggested that the inspection of the substation start at a location where one of the high voltage lines enters the switchyard and that is follows the path that the power takes through the substation. In general, the procedure that is suggested here is for equipment that is 220 kV and above and for sites that have experienced damaging ground motions. Clearly if time is very limited the suggested procedure

Table 6.1 Symbols used on a operating diagram

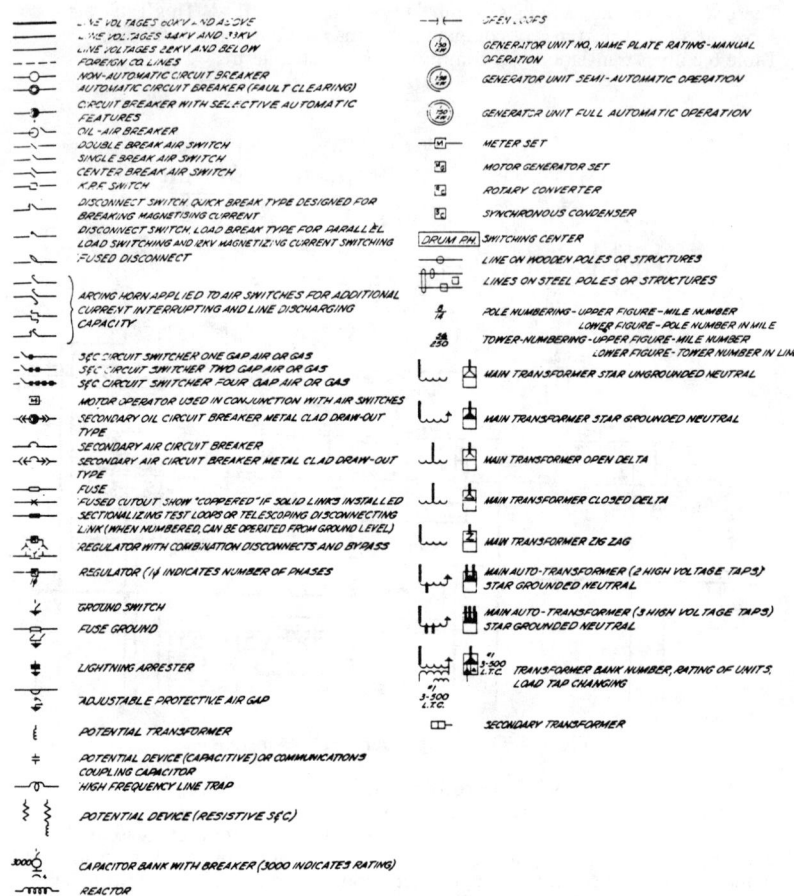

Table 6.2 Translation for Technical Power Terms

English	Spanish	French
Power Transmission Line	Linea de Transmision	Ligne de Transport (d'e'lectrcite')
Distribution Line	Linea de Distribucion	Ligne de Distribution (d'e'elctricite')
Oil Circuit Breaker	Interruptor de Aceite	Disjoncteur A' Huile
Air Blast Circuit Breaker	Interrupter de Aire	Disjoncteur A' Air Comprime'
Live Tank Circuit Breaker	?	Disjoncteur A' Faible Volume D'Huile
Dead Tank Circuit Breaker	?	Disjoncteur A' Bain D'Huile
Recloser	Actuador de Cierre	Disjoncteur A' Re'enclenchement
Fuse	Fusible	Coup-circuit (a'Fusible)
Lightning Arrester	Pararrayo	Parafoudre
Potential Transformer	Tranformador de Voltage o Potencial	Transformateur de Tension
Power Transformer	Transformador de Abasto	Transformateur de Puissance
Single Phase Transformer	Transformador Monofasico	Transformateur Monophase'
Three Phase Transformer	Transformado Trifasico	Transformateur Triphase'
Current Transformer	Transformado de Corriente	Transformateur de Courant
Air Disconnect Switch	Disconectivo de Aire	Sectionneur
Bus Support Structure	Estructura de Soporte para Barras de Conduccion	Charptente de Support de Barres
Synchronous Condenser	Condensado de Sincronizacion	Compensateur Synchrone
Capacitor Bank	Banco de Condensadores	Batterie de Condensateurs
Reactor	Reactor	Bodine D'Inductance
Line Trap	?	Circuit Bouchon
Post Insulator	Aislador de Poste	Colonne Isolante
Battery Room	Cuarto de Baterias	Salle Des Accumulateurs
Sudden Pressure Relay	Bobina de Presion Instantanea	Relais de Variation Brusque de Pression
Protective Relay	Bobina de Proteccion	Relais de Protection
Ceramic	Ceramica	Ce'ramique
Gasket	Empacadura	Joint D'e'tanche'ite'

Table 6.2 Translation for Technical Power Terms

English	Spanish	French
Bushing	Anillos de Proteccion	Traverse'e
Transformer Radiator	Radiator de Transformador	Radiateur de Transformateur
Liquid Storage Tank	Tanque de Almacenamiento de Liquido	Re'servoir Pour Storage de Liquide

may have to be abbreviated by just looking at damaged equipment. Your host may also have a preferred way of reviewing the damage with you. Pictures should be taken along the path that the power follows through the switchyard showing bus support structures, slack in bus between equipment, the equipment, and its anchorage.

When inspecting substations several safety procedures must be followed. Some utilities require that a long-sleeve, no-synthetic shirt be worn to prevent burns if equipment arcs. One should not point a finger or a metal object, such as a pen, within a substation, particularly if one is close to the equipment. Long conducting objects, such as a steel tape measure, should not be used in the substation. In general, it is best not to touch equipment.

As noted in the general procedure for conducting an investigation (See Section 5) an overall perspective of the site should be made. Some of this will be done in the process of walking through the site when looking at equipment. The following questions should be answered. What is the orientation of the site relative to magnetic north? What parts of the site are on cut or fill? By looking at the boundary of the site can an estimate of the depth of the cut or fill be made? If the site is on alluvium, inquire as to its depth. What is the depth of the water table? Are there any signs of soil or ground failure or faulting in or near the site? How would you characterize the soil type: rock, very firm soil, firm soil, soft soil? Does the site manager know if there was any special foundation preparations at the site? (If it is a new site this may be known, but in general this type of information is best obtained at the headquarters of the utility.

The list of equipment that follows has been organized in the order that if might be encountered when investigating a damaged site.

For each item of equipment, the following things should be checked as appropriate: slack in power connections to the equipment; the potential for interaction due to the relative motion between adjacent equipment items, between an equipment item and bus support, or due to dynamics of flexible bus. Follow the load path from the equipment into the supporting soil. Look at the base of bushings for signs of oil leaks, displacements of the gaskets, or displacement of bushings. Look at the interface between the equipment and its support structure. Look for cracked paint, displacements at interface, lose bolts, working of joints as indicated by scratches or burnishing of paint of galvanization. Look at the support structure and check connections, particularly slotted connections for working of connections. Look for signs of distress in the anchorage. This would include cracked or chipped

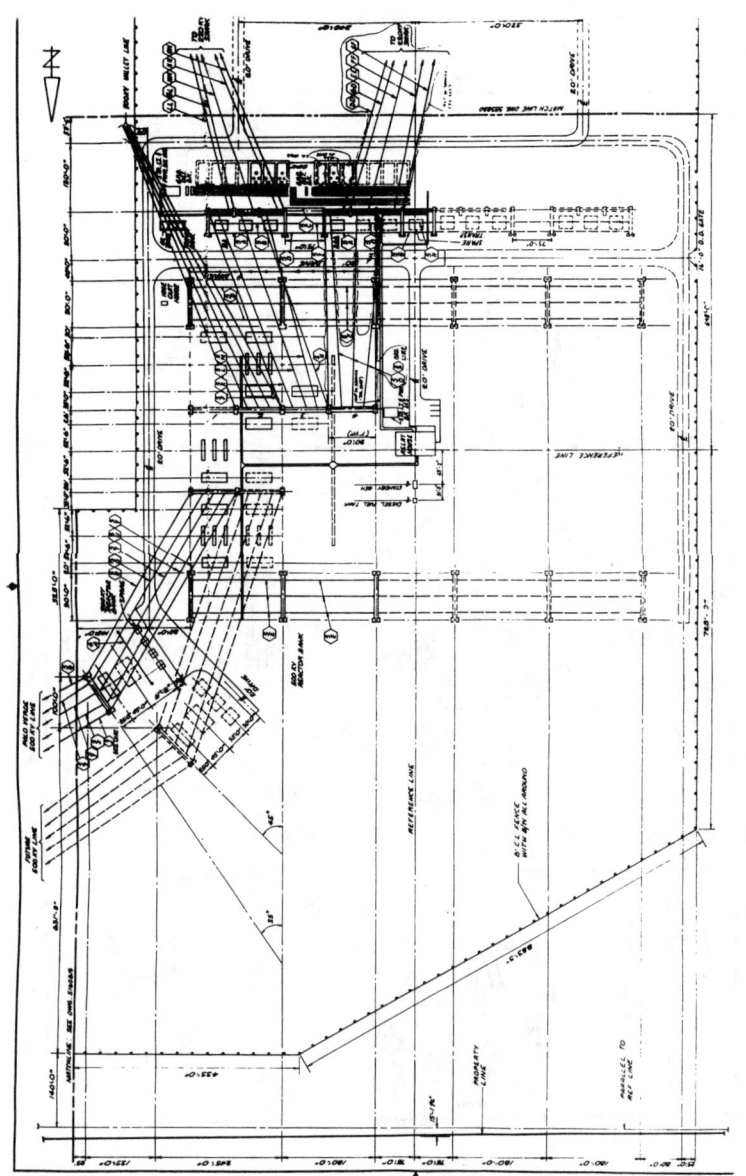

Fig. 6.6A Station layout diagram for 500 kV substation

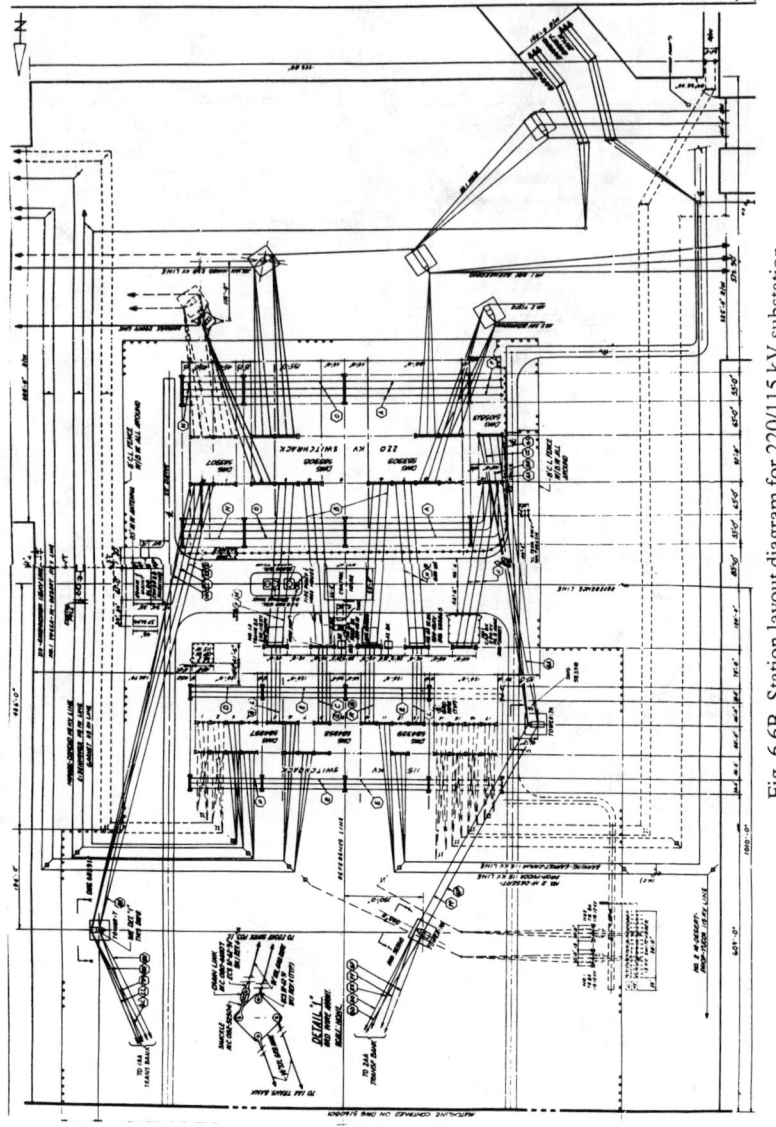

Fig. 6.6B Station layout diagram for 220/115 kV substation

paint in the equipment near the anchor points, displacement of the equipment on its base, looseness of anchorage indicating stretching or pull out of bolts, cracking or chipping of paint on bolts, chipping or spalling of concrete near anchor points, gaps around the foundation bed indicating compaction due to motion of the pad or settling of the pad. For damaged equipment the above information should be noted as well as the function of the equipment.

Current-Voltage Transformer

The current-voltage transformer (CVT) is often the first item of equipment connected to a transmission line after the line enters the substation. The CVT is used to measure voltage and its output is also used for system protection. The CVT is usually takes the form of a porcelain column attached to a box. Units located where lines enter the substation are often supported on on the same structure as a wave trap, as shown in Figure 6.7. They are also found adjacent to circuit breakers, in which case they typically have their own support structure.

Fig. 6.7 Current-Voltage transformer mounted adjacent to a wave trap

Line Trap

For higher voltages (>345 kV) it is common to have a wave trap on each phase. For lower voltages there may be a wave trap on only one of the phases. Wave traps are usually mounted in one of three ways: Suspended from flexible bus with its axis vertical (Fig. 6.8), supported on a single post insulator on top of a support structure with its axis vertical (Fig. 6.9), or supported by a post insulator at each end with its axis horizontal on top of a support structure (Fig. 6.7). Failure is usually of the post insulators that support the line trap on top of its support structure. These units, after they fail and the flexible bus supported line trap can also damage other equipment due to loads applied through bus connections.

Fig. 6.8 Suspended wave trap   Fig. 6.9 Vertically supported wave trap

Bus and Bus Supports

One type of bus system uses flexible aluminum cable and connections to equipment are done with vertical drops of flexible bus from the main overhead cables, Figure 6.10. The main bus lines are supported on dead-end type structures. While these systems will usually provide adequate slack, the dynamic response of the vertical drops or the dynamic response of the main lines may load equipment bushings connected to the bus.

Fig. 6.10   Aluminum cable used to connect a disconnect switch to overhead flexible bus

A second system is to support the bus, be it rigid or flexible, on vertical posts, usually a square tubular steel column, topped by a post insulator, Figure 6.11. Very often the bus is not provided with the necessary slack in connections to bushings. Bus supports may also be relatively flexible due to the use of thin base plates and a lack of gussets at the base plate-column connections. While it has not been common, flexible bus has been stripped off of most supports due to the failure of post insulators. On rigid bus, provision is made to accommodate thermal expansion of the bus. While this may provide some flexibility in connections, it will usually be inadequate for seismic response and is usually limited to a direction along the bus. Flexibility of bus connections perpendicular to the bus of often relatively rigid. Thus, vibration induce motion of the bus supports or of the equipment can apply significant loads to bushings.

Bus supports have also failed due to inadequate welds between the columns and their base plates, Figure 6.12. This same configuration is used to support lightning arresters, wave traps, current transformers, and other substation equipment. While the tubular steel supports are very strong and rigid, a thin base plate and a lack of gussets at the base can yield a rather flexible system. Earthquake induced motions in conjunction with inadequate slack in electrical connections can put large loads on ceramic members.

Fig. 6.11 Bus supported by post insulator on top of tubular steel column

Fig. 6.12 Poor weld penetration at column-base plate connection caused failure

Air Disconnect Switch

Each type of air disconnect switch has had problems associated with it. Pantographic types tend to be damaged when they are closed due to relative displacements between the ends of the switch which are typically connected to different structures. In this situation, each support structure will respond differently and the relative deflections will damage the switch.

Vertical-swing,"knife-blade" switches (Fig. 6.13) are more vulnerable when they are open in that the vertical link places large moments on its ceramic support member. This type of switch has been damaged when the bearing on the rotating member has been deformed.

Fig. 6.13 Vertical-swing air disconnect switch

Horizontal-swing switches (Fig. 6.14) appear to be less vulnerable than vertical action switches when each is open. When closed, some have had bearing supports at their bases deformed.

Most circuit breakers have air-disconnect switches adjacent to them that serve to isolate the circuit breaker. Live tanks circuit breakers, which have pressurized ceramic vertical members that support the live tank, have had their support columns fail with explosive force that have damaged near by disconnect switches.

Fig. 6.14 Horizontal-swing air-disconnect switch

Air-disconnect switches are often on flexible support structures so that there is a need for slack to adjacent equipment. In the case of circuit breakers, they are often very close to the air disconnect switch so that little or no slack is often provided.

Current and Potential Transformer

Current transformers consist of a large bushing on top of a box which is usually has its own support structure, Figure 6.15. They are often adjacent to circuit breakers and have experienced damage to the bushing, possibly due to interaction. Some current transformers are incorporated into circuit breakers, (Fig. 6.16). Damage is primarily to ceramic members.

Potential transformers have a form that is similar to current transformers. They are usually supported on their own support structure but in some cases are suspended from strain bus. Damage is usually to ceramic members and is associated with the dynamics of its support system or with interaction with adjacent equipment.

Circuit Breaker

Circuit breakers are probably the most vulnerable equipment item in a switchyard and they have exhibited several failure modes. There are many different types of circuit breakers and their vulnerability is closely related to their design and details of their installation.

The most rugged are oil circuit breakers, but these are primarily found at lower operating voltages. There main failure mode is poor anchorage that allow the units to slide and damage bushings. Friction clips are frequently used to anchor these devices. Three circuit breakers are often mounted on a single skid

Fig. 6.15   Stand-alone current transformer adjacent to a circuit breakers

(Fig. 6.17) which is then anchored with friction clips, Figure 6.18. Higher voltage oil circuit breakers (220 kV) are typically mounted as single units. Friction clips have two main problems. First, they tend to rotate about their anchor bolt allowing the anchored object to slide free. Second, the configuration puts twice the hold-down force on the bolt. One method to improve the seismic response of friction clips is to weld the clip to the anchored item.

Dead tank circuit breakers are the next most rugged type of circuit breaker. In this design the mechanical parts of the circuit breaker that interrupts the circuit are contained in a tank which is at ground potential. The tank is supported on a steel support structure near the ground. The bushings on these units would be the most vulnerable.

Live-tank circuit breakers have proved to be very vulnerable. The tank or interrupter head, which contains the interrupter mechanism, is at line voltage and is supported on tall ceramic columns. At higher voltages (220 kV and above) there is usually more that one interrupter head so that the circuit breaker will have from two to five columns. There are three types of support columns. One type is similar to a large post insulator, that is, the column is made up of one or more ceramic members, each with a metal flanges bonded at its ends, Figure 6.19. The lower flange will be bolted to the support structure or a gas

Fig. 6.16   Current transformer incorporated into circuit breaker interrupter head support column

Fig. 6.17 Oil circuit breaker anchored to pad with friction clips

Fig. 6.18   Friction clip welded to skip to prevent slipping and rotation

Fig. 6.19   Damaged air-blast circuit breaker with bolted flanged support columns

storage tank. The top flange is bolted to the interrupter. Failure modes of this type of circuit breaker are discussed in the case study at the end of the section.

A second type of circuit breaker column consists of ceramic cylinders that are separated by rubber type gaskets, Figure 6.20. They are held together by internal tendons, either wood or fiberglass, that are under tension. A third type of construction is a frame structure consisting of relatively slender ceramic members that are joined by aluminum framing in such a way that the interrupter head is electrically isolated from the ground, Figure 6.21. Note that one of these units failed and the interrupter head is missing. An aluminum casting that supported the head fractured, Figure 6.22. Typically air at pressures as high as 500 psi is used to extinguish or blow out the arc that is struck when a circuit breaker opens. Thus, these units will typically have a high pressure air supply connected to the live tank. In some cases the insulating columns are pressurized at about 60 psi with insulating gas.

Fig. 6.20 Air-blast circuit breaker with tendon-type support columns

Two other types of live tank circuit breakers were recently observed in the Soviet Armenian earthquake. These were 220 kV units. In one unit, which had two interrupter heads, the second head was placed above the first so that the resulting circuit breaker was very tall. In the U.S. units with two interrupter

Fig. 6.21 Braced-frame type live-tank circuit breaker

heads have each head supported by its own column. A second type had what appeared to be four interrupter heads connected in series and contained in ceramic insulators that resembles post insulators.

Several failure modes for live tank circuit breakers have been observed, the most common are associated with failure of the ceramic columns that support the interrupter heads. Flange mounted units tend to fail at the sand ring just above the lower flange. Pre-tensioned units tend to blow out the gaskets between members that make up the support column. These members also frequently fracture, Figure 6.23. Bushing gaskets on the interrupter heads have also blown. Another failure mode is the failure of ceramic members that carry high pressure air to the interrupter heads.

While most substations are designed with extra circuit breakers so that a single breaker can be quickly replaced, most earthquakes damage a large percentage of the circuit breakers at the site so that there are frequently lengthy delays to move in replacement breakers from other sites within the system. Even when spare parts are available, repairing a damaged circuit breaker that has lost its support column will require several days to a week.

Lightning Arrester (Surge Arrester)

Failure of lightning arresters have been quite common and usually occur at the sand ring at their lower flange. Lightning arrester can be mounted on its own support structure on transformers, as shown in Figure 6.24. In an emergency the system could be operated without the lightning arrester.

Fig. 6.22 Aluminum casting supporting the interrupter head fractured.

Transformer

Transformers have experienced several types of failures, many associated with inadequate installation practices. The main problem is that older units are inadequately anchored to rails and move or fall over, Figure 6.25. While many regions no longer rail-mount their transformers, they are frequently unanchored. As a result of movement of transformers there can be extensive damage to bushings, lighting arresters (Fig. 6.26), radiators, control cable connection (Fig. 6.27), bus connections and internal damage to the transformer. The most common failures are to lighting arresters (Fig. 6.24) and bushings due to inadequate slack in interconnections or the inadequate strength of ceramic members. In bushing assemblies that are installed under tension, gaskets which seal internal oil often are blown or develop leaks. This may require replacing the gasket and treatment of transformer oil, a lengthy process. Minor leaks to radiators are frequently observed but they are usually easily corrected

Fig. 6.23 Damage live-tank circuit breaker with tendon-type support columns

Fig. 6.24 Transformer-mounted lightning arrester. A failed unit is in the foreground.

Fig. 6.25 Transformers fallen off of their support rails

Fig. 6.26 Unanchored transformer slid causing failure of a lightning arrester

and do not disrupt service. Older transformers may use PCB as a dielectric material and their failure may result in leakage of hazardous materials that present a special problem.

Fig. 6.27 Movement of transformer damages control cables

Voltage Support Devices

Series or shunt capacitors are often used to stabilize system voltages from sudden changes in loads. The capacitor banks are usually supported on insulators to electrically isolate them. These supports or the structural faming used to hold the capacitors have frequently failed, Figure 6.28. Inductors and potential transformers associated with capacitor banks have also failed. Older capacitors may use PCB as a dielectric material and their failure may result in leakage of hazardous materials that present a special problem.

Substation Control Structure and Its Contents

In general control house structures have performed well. Problems are common with several items found frequently found in substation control structures. Station batteries have problems as noted earlier. Some types of protective relays can be activated by earthquake induced vibrations. This may cause some unwanted actions, but no damage to systems is known to have ever occurred as a result of spurious relay actions. After an earthquake, relays may have to be reset to resume operations. There have also been problems with panels and light fixtures mounted in suspended"T" bar ceiling falling or the entire ceiling can come down (Fig. 6.29). This is a hazard to personnel and to equipment in the control room.

Communications gear has experiences some problems, primarily due to the way that this type of equipment is installed. Communications racks often take the form used by the telephone industry, that is, they are 19" wide, 8" deep at

Fig. 6.28 Collapsed capacitor racks

Fig. 6.29 Fallen control room ceiling

the base, and relatively tall. Because of their narrow base, base anchorage is subjected to large overturning moments and the method of construction frequently yields a relatively flexible rack. The equipment in the rack can be quite heavy. There have been examples of circuit cards coming out of the rack, deflections in the rack damaging cable connections, and racks falling over.

Distribution (Feeder) Systems

The distribution system beyond the distribution substation is considered here. It consists of radial feeder lines, typically operating at 4 kV, 12 kV, 16 kV, or 21 kV, reclosers, manually operated switches, fuses, and pole, platform or vault mounted transformers. Figure 6.30 shows a schematic diagram of such a system.

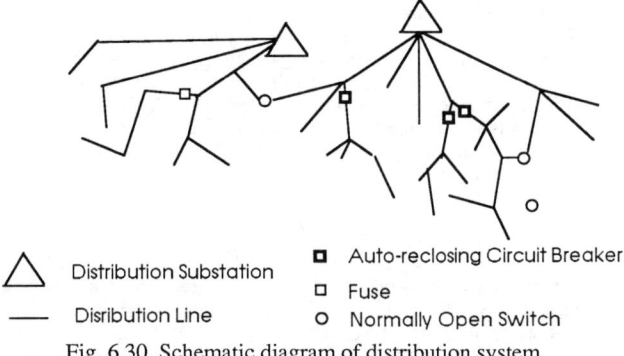

Fig. 6.30 Schematic diagram of distribution system

Two types of failures to the distribution system have been common 1) toppling of unanchored distribution transformers and 2) burn down of feeder and service lines. Pole-mounted transformers which hook over the cross bar rather than being bolted to their support have frequently become unhooked and have fallen. Platform mounted transformers, often three units supported on a platform formed by two 4" x 4" between two poles, are often unanchored to the platform and unrestrained. While a limited service area will be disrupted by each such failure, in the 1952 Kern County earthquake over 800 units failed and a large number of customers were left without power. Restoration of these units is labor intensive and with such large numbers of failures restoration times will be long.

Failure of feeder and service lines has also been common. Outside of the U.S. it is common to support distribution lines from existing structures, often in old communities where there is no room in the narrow streets to install poles. Lines are often lost when the structure fails. In recent U.S. earthquakes, there have been many cases of lines coming in contact and burning down, or occasionally wrapping around each other. Downed lines are a serous safety hazard. They have started fires, disrupted traffic when they fall across roadways, and have been a large load on the emergency response community. In recent California earthquakes, a large percentage of emergency response calls were associated with downed power lines and a large percentage of the earthquake related fires were caused by downed lines.

FACILITIES AND EQUIPMENT WITH GOOD EARTHQUAKE PERFORMANCE

High Voltage Transmission Lines

With the exception of a few direct current transmission lines and associated facilities, power is transmitted over three sets of conductors, each carrying power at 60 Hz (50 Hz in most of the rest of the world) with the voltage on each set of conductors leading or lagging the other by 120 degrees.

These lines and their support structures have proved to be very earthquake resistant. There are several reasons for this. First, the lines and their support structures must resist significant wind loads and broken wire conditions so that the towers must be strong. Second, the tower has to support relatively light masses attached directly to the tower. For many configurations, low natural frequencies of the transmission lines decouples their mass from transverse loads. Also, the natural frequencies of these systems is typically well below peak seismic inputs. Finally, the slack in lines can accommodate the relative motion between towers. The primary reason for the loss of transmission towers from earthquakes is ground failure that damages the foundation of a tower or an earthquake induced landslide that sweeps the tower away.

General Power System Equipment

Power systems, particularly power generating stations, contain a wide variety of equipment including control panels, distribution panels, motor control centers, pumps, motors, metal-clad switchgear, valves, etc. In general, this equipment is small and is shipped as pre-assembled units. Typically, shipping loads will exceed seismic loads so that if the equipment is installed with adequate anchorage, it will not be damaged. While the equipment may not be damaged, it may not function as intended. In particular, relays may experience a change in state as a result of the earthquake induced vibrations so some equipment may not function as intended. In most cases, the fact that a motor turns on or off and has to be reset does not cause any damage or disruption. One should inquire if such malfunctions have occurred and if they caused any problems.

CASE STUDY : Investigation of Tejon Ranch Earthquake/Edmonston Switchyard

The investigation of the switchyard associated with the Edmonston Pumping plant is used as a case study of an earthquake investigation. The damage occurred in the 1988 Tejon Ranch earthquake, which had a Richter Magnitude of 5.3. The main thrust of the case study is to illustrate the determination of failure modes and factors that contributed to damage. The substation was constructed in 1968 and was designed to 0.5g. The equipment, however, was designed to 0.2g static, the common practice at the time.

The most significant damage was to circuit breakers in the 230 kV switchyard. Four different types of failures were observed. A GE ATB-7 circuit breaker is shown in Figure 6.31 A schematic of the circuit breaker is shown in Figure 6.32 and a letter

Fig. 6.31 Live-tank circuit breaker (3000 amp.)

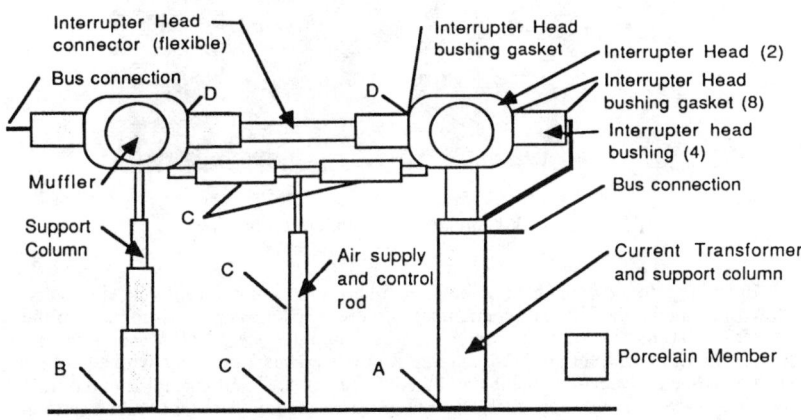

Fig. 6.32 Schematic diagram of a GE ATB-7 live-tank circuit breaker

indicates the location of each type of damage. Shaded members are made of porcelain. The circuit breaker has two interrupter heads, each supported by a porcelain column. The right support column, referred to as the current transformer support column, contains the leads of a current transformer and is pressurized with SF6 insulating gas to 60 psi. Five of these columns failed at their base (indicated by the letter A in the schematics). Two failed explosively, so that fragments damaged other porcelain members. Figure 6.33 shows a circuit breaker with the shattered current transformer support column missing and a U-shaped link between the interrupter heads pulled so that it is nearly straight. The left support column, which serves only to supports its interrupter head, has a smaller diameter than the current transformer support column and is not pressurized. One of these support columns failed at its base (indicated by a B in the schematics). Seven air supply systems failed at one or more locations (indicated by a letter C in the schematics). These systems consist of several porcelain members connected by pipe and have a "T" configuration. These columns are also pressurized with SF6. Within the column is a fiberglass tube that contains the high pressure (500 psi) air used to extinguish the arc created when the circuit breaker opens. Within the tube is a control rod that activates the circuit breaker. Figure 6.33 shows a unit where both arm of the T are broken and a center portion of the stem of the T is missing. The components of the system including the internal fiberglass tube and control rod can be seen in a damaged assembly shown on the ground in Figure 6.34.

Fig. 6.33 Damaged live-tank circuit breaker

In addition to the damaged porcelain members, eighteen interrupter head gaskets blew (indicated by a D in the schematics). Figure 6.35 shows a close-up view of the gasket protruding from the interrupter head-bushing interface. Each phase of the circuit breaker has 2 interrupter heads. Each interrupter head has 2 bushings with a gasket at each end, one between the bushing and the interrupter head and one at the end of the bushing to seal the bus connection. The only gaskets to fail were those adjacent to the interrupter heads on the bushings between the interrupter heads. These bushings are connected to each other by heavy U-shaped sheets to provide some flexibility between the two interrupter heads.

Fig. 6.34 Damaged air supply system showing components

Fig. 6.35 Blown interrupter head-bushing gasket

One support column on a disconnect switch failed at its base. Its position, indicated by an asterisk and the distribution of other types of damage are shown on the Schematic of Circuit Breaker Layout, Figure 6.36.

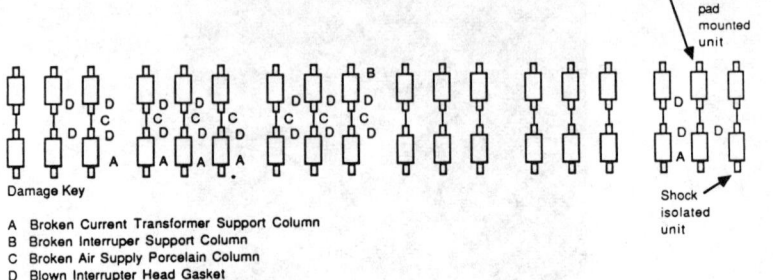

Damage Key

A Broken Current Transformer Support Column
B Broken Interruper Support Column
C Broken Air Supply Porcelain Column
D Blown Interrupter Head Gasket
* Broken Disconnect Switch Support Column
   Connected to this Circuit Breaker

Fig. 6.36 Schematic diagram of circuit breaker site layout

Other Observations

The shock isolated circuit breaker (1 phase) was not damaged, however, 7 of the other 15 phases that were not isolated were also undamaged, so no conclusions can be drawn about the performance of the isolated system. The circuit breaker that was on elastomer pads did experience some damage, but the pads probably reduce only very high frequency vibrations.

All equipment within the switchyard was well anchored and there were no signs of "working" or distress in any of the anchorage or equipment supports. Conductor slack between equipment in the switchyard varied. Following the circuit from the power plant side of the switchyard toward the input lines, ample slack was provided between the strain bus and the disconnect switch, Figure 6.14. Almost no slack was provided between disconnect switches and circuit breakers, Figure 6.31. This figure also shows a large muffler hanging off of the side of each interrupter head. The muffler would give the mass of the interrupter head a large eccentricity. Figure 6.37 shows the U-shaped connection between the two interrupter heads. This connection is made of 8 laminated aluminum sheets, each about 1/16" thick, 4" wide with the depth of the U about 6". For small deflections this would give a stiffness of about 45 pounds/in. in a direction along a line connecting the two heads. It would be much stiffer in the perpendicular direction. On the 3000 amp circuit breaker, two U's are connected in parallel. The bar which forms the connection between the interrupter head bushing and the input to the current transformer contained in the current transformer support column is shown at the right side of Fig. 6.31. The connection from the circuit breaker to the disconnect switch does have some flexibility. This figure also shows the base isolated circuit breaker in the foreground. Note that the slack at this connection on the base isolated circuit breaker is the same as on the fixed base units in the background. Extra

Fig. 6.37 Connection between circuit breaker interrupter heads

slack was provided to the other connection of the base isolated circuit breaker. No slack was provided between the two disconnect switches as shown in Figure 6.38.

The blown interrupter head gaskets, cracked current transformer columns and air supply columns caused a loss of insulating gas and air pressure so that the circuit breakers could not be opened. Evidently circuit breakers at the SCE Pastoria substation relayed, isolating the damaged equipment. Bulk power in the area is supplied by both PG&E and SCE. The need to reset digital clocks that do not have battery backup indicated that there was a momentary loss of power on the SCE system.

Two strong motion seismographs were located at the site, one in the transformer yard adjacent to the plant, and one in the coupling gallery low in the pumping plant. The transformer yard unit did not operate due to a jammed trigger. The initial estimate of the peak acceleration recorded in the plant was 0.08g. This supports the physical indications that the site did not experience large accelerations. The duration of the strongest shaking lasted about 1 second with the horizontal components showing sinusoidal like character. One had a frequency of about 3 Hz. and the other about 4.5 Hz. These motions may have been influenced by the response of the structure in which the recorder was installed. Motions at the seismograph location may have been attenuated by about 25% compared to a free-field location due to the effect of the character of the power plant structure and its embedment. Parts of the switchyard may have seen some what larger motions because some of it is located on fill.

Figure 6.39 shows that the switchyard shelf was extended with fill into a gully. While construction drawings would have to be checked to determine the depth of fill, if any, at each circuit breaker location, it would appear from the figure that there is little or no fill at each end of the line of circuit breakers and the maximum depth of fill occurs at about 1/3 of the way from the right end of the installation. It should be noted that the damage is, in general, the least where the fill is the deepest.

Fig. 6.38 Connection between disconnect switches

Fig. 6.39 Switchyard showing fill in gully

It has been a common practice in earthquake damage reports to refer to damaged equipment by its generic name, such as an "air blast circuit breaker" or a "live or dead tank circuit breaker." The configuration and seismic performance of units from the same manufacturer can be quite different. For example, the GE ATB-7 circuit breakers at the Edmonston site are significantly different and have different failure modes than the GE ATB-5 circuit breakers observed in the Whittier Narrows earthquake. In the ATB-5, the vertical columns are segmented, separated by gaskets and held together by internal tendons, while in the ATB-7 columns are flange mounted. The ATB-5s had leaking gaskets between elements of the vertical columns while their bushing gaskets did not leak. Even at the Edmonston site there were two types of circuit breakers, both designated as ATB-7 circuit breakers. In one unit the interrupter heads weighed 3600 pounds while in the other the weight was 2600 pounds. The only difference in the model numbers of these units was one digit in the 22 that describe the units.

Speculation on the Causes of Damage

Three types of circuit breaker damage will be considered: 1) the cause of the blowing of the interrupter head bushing gaskets, 2) the failure of the porcelain interrupter head support columns and, 3) the failure of the pressurized air system. It should be noted that these circuit breakers are of about 1970 vintage so that their seismic requirements would have been a horizontal static 0.2 g.

Causes of the Blown Interrupter Head Bushing Gaskets

The gaskets and the two bushings on each interrupter head are identical and are installed in the same way. Yet all interrupter head bushing gaskets that failed were on the inside bushings (those which joined the two interrupter heads) rather than the outside bushings connected to the bus or current transformer connection. Of the 18 blown bushing gaskets, 11 were on circuit breakers which had no support column damage. There were also blown gaskets on circuit breakers with undamaged air supply systems. This pattern of failures suggests that there were significant relative motions between the two interrupter heads, even when the support columns did not fail. The porcelain part of the interrupter head support columns and the member to which they are anchored appear to be very rigid. However, the coupling of the porcelain to its flange can introduce significant flexibility at this joint. The natural frequencies and damping of this equipment has been found to be quite low, with frequencies about 2 Hz and damping about 1% of critical. This would appear to be the source of the motion that caused the gaskets to blow. The U-shaped connecting member between the heads does provide some flexibility but it can still transmit loads when the heads move relative to each other.

The outside bushing on the interrupter head on top of the current transformer support column is connected to the column so motion of the column would not introduce any load on the bushing or the gaskets. The outside bushing on the other interrupter head is connected to a disconnect switch with almost no slack (Fig. 6.32). The fact that its gaskets did not blow suggests that the failure was not due to the lack of slack. However, one of these circuit breaker columns and one disconnect switch column did fail. It should be noted that for each of these cases, the inside gaskets failed, so if they failed first, the internal pressure could have be released saving the outside gasket from blowing. The lack of slack was probably the cause of these failures.

Failure of Interrupter Head Current Transformer Support Columns

There are two possible primary failure modes; one associated with the mass of the interrupter heads, and the other from the pull on conductors associated with the relative deflections between connections. Some evidence supports each failure mode. Also, nonlinear effects such as rapid changes in the stiffness of connections with change in relative deflection and impacting of parts that have broken can have a profound influence on the failure of porcelain members.

One explanation for the column failures is that the inertial loads imposed by the earthquake exceeded the strength of the current transformer support columns. Since the current transformer support column is larger in diameter than the other support column, it should be more rigid and thus draw more of the inertia loads acting on the interrupter heads. The relative strength of the two columns is not known. As discussed earlier, the blown interrupter head bushing gaskets indicate that the interrupter heads moved relative to each other. While the explanation for this motion was flexibility at the flange-porcelain interface, this flexibility would, in general, reduce the stress at the root of the porcelain column. It could be that the more rigid current transformer column was loaded through the U-shaped coupling from the other support column.

There is contradictory evidence, however. Most interrupter heads weigh about 2600 pounds, but the unit on the right and the fourth unit from the right in Figure 6.38 weigh 3600 pounds. Structurally both types of circuit breakers are the same. Only one support column on the 6 phases with heavy interrupter heads failed. One would expect that the inertial loads would be more severe for the heavier units, when in fact they suffered less damage. Also, the only non-current transformer support column that failed was on a circuit breaker that had the current transformer support column undamaged. The fact that the support columns on the five other circuit breakers with failed current transformers support did not fail suggests that the failures were due to lack of slack rather than inertial forces. If the failures were due to the inertial forces, the single support column that remained after the first column failed would have to support additional lateral load from the heads with the failed columns. One would expect that this would cause the single remaining column to fail unless the character of the ground motion was such that there was no significant motion after the first support columns on each unit failed. Finally, the unit with the heavier head would have had a lower natural frequency which may have reduced the effect of the input ground motion.

It was noted earlier that the slack between the circuit breaker bus connection on the current transformer support column and the disconnect switch was marginal. Thus, a relative deflection between the circuit breaker and the disconnect switch would put a load on the current transformer column (five of which failed). While the circuit breaker and disconnect switch support structures were of very heavy design, the tubular steel columns did not have gussets at their base plates. While these structures had great strength, they may have been flexible enough to contribute to the damage the circuit breaker support columns.

The performance of the base isolated unit also raises some questions, even though it was not damaged. The connection between the current transformer support column and the disconnect switch was the same on the base isolated unit as on all other units. The relative deflections of the base isolated unit would be expected to be much larger than the fixed base units so that larger loads should have been applied to the support column due to the lack of slack, yet it did not fail.

The most probable cause for the failure of the current transformer interrupter head support columns is from the dynamic response of the interrupter heads with possible contributions from the eccentricity of the mufflers and nonlinear effects.

It should be noted that the variability of the strength of porcelain is much larger than most structural materials so some failures may occur at relatively low loads. This may distort any pattern of failures and make it difficult to identify failure modes. Also, the role of the location and depth of fill under the switchyard is unknown. Clearly the above conclusions must be considered very speculative in light of the limited and contradictory data .

Failure of the Pressurized Air System

Seven of the 16 air supply systems failed. The system is configured in the shape of a "T" with porcelain members in the stem and arms (Fig. 6.37). The porcelain members are relatively slender but support no additional weight. The base of the "T" is anchored to the circuit breaker support and the end of each arm is anchored to one of the interrupter heads. The rigidity of these connections is not known. Failures of the stem and arms of the "T" were common. While 2 units that failed were on circuit breakers that had no broken interrupter head support columns, each failure was accompanied by blown interrupter head bushing gaskets. The blown gaskets would indicate that there was significant relative deflections between the two interrupter heads as discussed above. This would account for the damaged arms and the failure at the base of the "T". While some air supply columns failed at their bases, most had damaged porcelain members near the center of the column. This would suggest a failure due to the vibration of a free-free beam. Some of these units may have been damaged by flying debris.

Several fiberglass tubes that contain the high pressure air supply failed. If these failed before their porcelain support column, high pressure air would be released into the gas insulated bushings. This could account for the blown gaskets and the current transformer support columns that exploded. However, since the fiberglass tube is much more flexible than the porcelain bushing which surrounds it, it is likely that the porcelain failed before the fiberglass, allowing the high pressure air to vent to the atmosphere. Also, there were two circuit breaker phases that had blown gaskets with the air supply undamaged. The explosion of two current transformer support columns could have been caused by the 60 psi gas, which each contained, after seismic loads cracked the columns.

One would expect from the needs to assemble and maintain the equipment that the "T" could be supported at each end in a horizontal position, thus subjecting it to a 1g load. Even with dynamic amplification it is unlikely that acceleration of this magnitude would have been experienced without nonlinear effects.

POWER SYSTEM CHECK LISTS

Anchorage Check List

___ Are the anchor bolts cast in place or expansion anchors?
___ Can you identify the type or manufacturer of the anchor?
___ What is their length of embedment?
___ What is their diameter?
___ How did they fail?
___ Did they pull out of concrete?
___ Is the concrete cracked?
___ Did fracture cones develop in the concrete?
___ Did the bolts stretch?
___ Did the bolts break?
___ Is there any indication that they were installed incorrectly?
___ What were the standards, if they existed, when the equipment was installed?
___ How many bolts were there and how were they laid out?
___ Did the bolt pass through a structural member in the equipment framing?
___ Are there signs of distress in the equipment in the region around the anchor bolt: cracked or chipped paint, deformation of metal?
___ Does the equipment introduce a prying action to the bolt?
___ Is the bolt hole appropriate to the bolt diameter?
___ Does the load path from the equipment frame to the bolt or weld introduce flexibility in the anchorage system?
___ What are the sources of loading on the anchorage: equipment weight, height of center of gravity, dimensions of the base of the equipment?
___ Were there loads applied through interconnections to adjacent equipment?
___ Has the base of the equipment moved on its footing?
___ Is there a gap around the footing or equipment pedestal indicating differential movement?

Power Equipment Check List

___ What slack in power connections to adjacent equipment equipment?
___ Is there a potential for interaction due to the relative motion between adjacent equipment items, between an equipment item and bus support, or due to dynamics of flexible bus?
___ Have you followed the load path from the equipment into the supporting soil?
___ Have you looked at the base of bushings for signs of oil leaks, displacements of the gaskets, or displacement of bushings?
___ Have you looked at the interface between the equipment and its support structure? Check for cracked paint, displacements at interface, lose bolts, working of joints as indicated by scratches or burnishing of paint or galvanization.
___ Have you looked at the support structure? Check connections, particularly slotted connections for working of connections.
___ Review the Anchorage Check List.

Power Plant Check List

___ Review Site Check List
___ Inquire about damaged equipment.
___ Look for interaction problems between the boiler support structure and the boiler.
___ Look for interaction problems between the turbine pedestal and the powerhouse operating floor.
___ Inquire if the unit went off line. If so, determine why.
___ Are there any indications of turbine bearing damage?
___ Does there appear to be steam coming from the stack indicating boiler tube damage?
___ Generally, how is equipment anchored?
___ Check for damage to coal handling equipment.
___ Check station batteries.
___ Were sudden pressure relays in transformers activated?
___ Did any protective relays change state? Which ones?
___ Were any relays reset after the earthquake to resume operations?
___ Was there a loss of power on any lines into or out of the station?
___ If there was any disruptions, what was the cause and what was the duration?
___ If there was a suspended ceiling, did any of the panels fall?
___ Did anything fall from desks, tables or shelves in the substation?
___ Were there any disruption in communications? If so, what types of communications are used and which were effected?
___ Have personnel that were on the site at the time of the earthquake describe the earthquake and their actions after the earthquake.
___ Are the personnel aware of any other effects that the earthquake had on the power system?

Substation Check List

___ Review Site Check List
___ Check for damaged equipment.
___ For vulnerable equipment (circuit breakers, lightning arresters, transformers, current and potential transformers, capacitor racks, and line traps) review Power Equipment Check List.
___ Check station batteries.
___ Were sudden pressure relays in transformers activated?
___ Did any protective relays change state? Which ones?
___ Were any relays reset after the earthquake to resume operations?
___ Was there a loss of power on any lines into or out of the station?
___ If there was any disruptions, what was the cause and what was the duration?
___ If there was a suspended ceiling, did any of the panels fall?
___ Did anything fall from desks, tables or shelves in the substation?
___ Were there any disruption in communications? If so, what types of communications are used and which were effected?
___ Have personnel that were on the site at the time of the earthquake describe the earthquake and their actions after the earthquake.
___ Are the personnel aware of any other effects that the earthquake had on the power system?

7. Water Systems

    System Configuration
    Sources of Water
    System Facilities
    Emergency Response Plan
    Operations
    What to Look For
    Common Failures
    Storage Facilities
    Water System Check Lists

7. WATER SYSTEMS

SYSTEM CONFIGURATION

A water system is a interconnected network of pipes, pumping, storage and treatment facilities supplying water for domestic, industrial, agriculture and fire suppression. Water systems consist of one or more pressure zones supplying water in the pressure range of 35 psi to 125 psi. These pressure zones are identified by a number representing the maximum pressure in pounds per square inch (psi) or the elevation above sea level in feet, the elevation which the water would theoretically stand under static conditions.

The importance of having a functioning water system after natural disaster, including earthquakes, is as follows:

1. For fire suppression.
2. For emergency (Medical, Public Safety and Emergency Operating Centers) facilities involved in the recovery after the disaster.
3. For restoration to individuals water for bathing, cooking and sanitary purposes.
4. For restoration of business and industry for the economy and employment.
5. Minimizing the damage from flooding to other facilities.

SOURCES OF WATER

The normal sources of water are diversions from rivers and streams, groundwater basins and springs. Some of these sources are located long distances from the service areas which require conveyance facilities, such as, aqueducts, consisting of pipes, conduits, canals and tunnels. Gravity aqueducts may have hydroelectric power plants, while non-gravity aqueducts require pumping plants.

SYSTEM FACILITIES

Conveyance Facilities

Conveyance facilities consist of transmission lines (trunk lines), distribution mains, conduits, canals, tunnels, etc. Most pipelines and conduits are buried, however, there are some that are supported by piers or bridge structures. Pipe materials are cast or ductile iron, steel, concrete, asbestos cement, or plastic, connected by joint systems that are either mechanical, welded, or caulked. Most materials used in water transmission systems are brittle and susceptible to fracture when subjected to deformation.

Pumping Facilities

Pumping facilities consist of booster pumping stations (pumping plants) to increase pressure or raise water to higher elevations and groundwater pumping stations (wells) to extract water from the ground-water aquifers. The facilities are usually enclosed in a building or outdoors at ground level. Some wells have pumps and motors (submersible pumps) at the bottom of the well. A power substation is a part of the facility.

Storage Facilities

Storage facilities consist of tanks, surface reservoirs and groundwater basins. The tanks are made of steel, wood, concrete, plastic, etc. and can be buried, ground level or elevated. Reservoirs are created by a dam to impound water from rivers and streams or may be an excavated earth basin with a lining. Most small open reservoirs are covered by wood, concrete, plastic, or metal roofing systems or floating covers.

Pressure Reducing Facilities (Regulators)

Pressure reducing facilities consist of a single regulator valve or a group of regulator valves to reduce the pressure from high to low pressure to protect the downstream piping from bursting. Also at the same location or nearby is a relief valve facility, which discharges into a drainage channel. This facility is for the purpose of protecting the system from high pressure in case of a failure of the primary regulators. Most of these facilities are located in buried vaults.

Treatment Facilities

Treatment facilities consists of chemical injection into the water supply for disinfection (chlorine plants or chlorination stations), for dental health maintenance (fluoridation), or complete water treatment plants (predisinfection, coagulation, sedimentation, filtration and post chlorination). Most chemical injection facilities consist of an enclosed building for mixing and storage of chemicals and an injection chamber (vault) at the pipeline or conduit. Full treatment plants consist of piping, basins, tanks, and pumps with assoicated mechanical and electrical equipment for water treatment.

Service Facilities

Service facilities consists of service connections, customers private property pipes, meters, fire hydrants, blow offs and backflow prevention devices. The service connection joins the distribution main to the meter and is made of copper, plastic, galvanized iron, lead, etc. Fire hydrants are usually made of cast iron. The back flow prevention facility is one or more back flow valves (cast iron) which prevents an exterior source of pollution from entering the water system when there is a loss of pressure.

Control Center

Equipment in the control center varies from simple recording charts that record flows, pressures, water levels, etc. to very sophisticated computer data acquisition systems with remote and/or computer aided operations system called SCADA (Supervisory Control and Data Acquisition). The center is usually located at the water system administrative center and is very dependant on the communication systems (wire, radio, microwave, etc.) between the remote operating unit and the center. The center usually responds to day to day customer emergencies and dispatches crews and communicates hydraulic operational changes.

Service Yards

Service yards normally contain material storage areas, shops, and administrative and maintenance facilities. It is a location for personnel to report to after a disaster to meet with their supervisors for assignments in the recovery operation. From the Service Yard they can be dispatched with the proper repair material and construction and portable operating equipment. The service yard consists of a number of buildings, storage sheds, and fueling facilities.

EMERGENCY RESPONSE PLAN

Most utilities respond to emergencies 24-hours a day every day. However, they have a Emergency Response Plan for major disaster, which includes water supply interconnections (permanent and temporary) with other systems and formal and informal mutual aid agreements with other utilities through state offices of emergency services.

OPERATIONS

A water system supplies water by gravity or pressure on demand. Pumping station and storage facilities operate automatically based on tank and water levels. Storage facilities reduce the peak demand on the system. Treatment facilities maintain water quality to meet or exceed Federal and State water quality standards.

WHAT TO LOOK FOR

In addition to specific water system failures, information that can be used to determine the number of pipeline failures per unit length for each of the various types of pipes in areas of damage is important for planning purposes. A map of the water system at the time of the earthquake and a delineation of earthquake intensity in the service area is very important to put specific failures into perspective.

1. Uncontrolled surface water leaks
2. Changes in water flow and/or pressure
3. Damaged waterworks structures and enclosures
4. Ground displacement
5. Evidence of ground shaking
6. Landslides
7. Ground subsidence or settlement
8. Liquefaction
9. Bursting of pipes and fittings from internal high surge pressures
10. Facility's ability to function as to quality and quantity
11. Restorability of function-minor repair, moderate repair or reconstruction
12. Potential sources of pollution
13. Foundation or bedding of facilities
14. Good performance of seismically designed and constructed facilities

COMMON FAILURES

Conveyance Facilities

1. Pulled or compressed and split mechanical, rubber gasket or caulked pipe joints
2. Pipe barrel broken circumferentially in bending
3. Bursted pipe or fittings from pressure surges
4. Broken pipe, valves and fittings
5. Cracked walls in canals, conduits and tunnels
6. Damage to pipe and other metal fixtures weakened by corrosion
7. Canal and conduit embankment slides
8. Cracked support cradles for elevated pipes and conduits
9. Flow restriction (plugging)-intrusion for foreign objects in the system
10. Joints failed by rotation

Pumping Facilities

1. Loss of power supply
2. Motors and pumps out of alignment
3. Shifting or toppling of mechanical/electrical equipment
4. Shifting or toppling of control cabinets
5. Emergency battery rack shifting or collapse which interrupts emergency generators or instrumentation systems
6. Building or enclosure damage or collapse which puts equipment out of service
7. Emergency power supply failure
8. Fuel storage tanks and piping leaks
9. Well casing mis-alignment; damage to connecting system
10. Potential contamination when water well is adjacent to broken sewer lines

STORAGE FACILITIES

Tanks (See Section 12)

1. Water level at time of disaster
2. Buckled, cracked and ruptured walls
3. Damaged or collapsed roof structure
4. Ruptured or cracked inlet-outlet piping
5. Vertical pounding on footing of anchored or unanchored tanks
6. Elevated tank support structure with bent or sheared bracing, broken bolts and buckled columns

Reservoirs

1. Cracks or settlement in earth or concrete dams and abutments
2. Surface slumps or sand boils
3. Bulging at the toe of the dam
4. Damage to spillway, inlet or outlet structures
5. Changes in quantity of flow and color of water in seepage monitoring devices
6. Changes in the water level or pressure in foundation or embankment piezometers

NOTE: State or federal dam safety authorities will make a detail investigation

Groundwater Basins

1. Damage to diversion and spreading ground structures
2. Damage to basin levees
3. Entry of sources of possible contamination
4. Change in hydraulic grade-elevation which changes flow direction

Pressure Reducing and Relief Facilities

1. See applicable failures under Pumping Facilities
2. Cracking of walls and roof of vault
3. Flooding of vault
4. Damage to electrical, mechanical or hydraulic controls

Treatment Facilities

1. See applicable failures under Conveyance and Pumping Facilities and wastewater treatment (Section 8.0 Sewerage Systems)
2. Clear water and chemical storage facilities-see applicable failures under Storage Facilities
3. Structural failure of buildings and enclosures
4. Chlorine gas leaks

Service

1. Broken service pipe and fittings at connection to main and meter
2. Broken meters
3. Meters plugged with foreign material
4. Broken fire hydrants or connecting piping
5. Broken back flow devices
6. Damaged emergency blow-offs

Control Center

1. See applicable failures under Pumping Facilities
2. Communication lines outage
3. Structural failure of building
4. Telephone system overload

Service Yards

1. Structural failure of buildings and storage sheds containing materials, construction and portable operating equipment
2. Structural failure of buildings housing administrative and control center operations

WATER SYSTEM CHECK LISTS

Conveyance Facilities

__ Description, get system map that can be keyed to a topo map
__ Get system statistics: lengths of a given diameter and material, ages
__ Operational status
__ Foundation or bedding
__ Type of materials
__ Joint system
__ Type of damage, leak, break(with/without flow), at coupling, bend, "T", corrosion, can the break be related to setting, topography, soil conditions, installation, etc?
__ Age
__ For each failure, locate on map and determine the above information
__ Get damage statistic for the system
__ Does the utility have any information on damage to service connections?
__ What factors in to area contribute to corrosion: Resistivity of soil (low contributes to corrosion), ground current sources in the area.
__ For damage sites visited, is there evidence of ground deformations?

Pumping Facilities

__ Description
__ Operational status
__ Foundation
__ Building or enclosure status
__ Evidence of liquefaction or settlement around building foundation
__ Power supply
__ Emergency power supply: starting, cooling system, vibration isolation, fuel storage
__ Piping
__ Emergency battery rack
__ Well equipment
__ Potential sources of pollution
__ Controls

Tanks (See Tank check list - Section 12)

Reservoirs

__ Description
__ Height, diameter and capacity
__ Operational status
__ Determine water level at time of event
__ Note if Federal or State dams authorities making detailed investigation
__ Upstream and downstream surface
__ Right and left abutments
__ Spillway, inlet and outlet
__ Hydroelectric facilities

Customer Damage

__ Damage to service connections, statistics and details
__ Damage to distribution system within the user facility, causes and impacts

Groundwater Basins

__ Description
__ Operational status
__ Flow structures
__ Levees
__ Direction of flow
__ Potential sources of pollution

Pressure Reducing and Relief Facilities

__ Description
__ Operational status
__ Vault condition
__ Mechanical, electrical and hydraulic controls
__ See check list under Pumping Facilities

Treatment Facilities

__ Description
__ Operational status
__ Foundation
__ See check list under Pumping Facilities
__ See check list under Sewerage Systems (Chapter 8.0) Wastewater

Treatment Plants Services

__ Description
__ Operational status
__ Damage
__ Plugging
__ Bursting

Control Center

__ Description
__ Operational status
__ Building status
__ Communication lines

Service Yards

__ Description
__ Operational status
__ Building status

8. Sewerage Systems

   Sewage System Configuration, Facilities and Operation
   Description of a Typical Sewerage System
   Essential and Non-Essential Functions of a Sewer
      System
   Dangers in an Earthquake Damaged Sewerage System
   Sewerage System Equipment and Their Functions
   Sources of Sewerage System Damage Information
   Seismic Performance of Sewerage System Facilities and
      Equipment
   Guide for Investigation Specific Facilities and
      Equipment
   Investigating Treatment Plants and Pump Stations
   Field Investigation of Failed Equipment
   Investigating Treatment Plant Structures
   Case Study of an Earthquake Investigation of a
      Collection System
   Sewage System Check Lists

8. SEWAGE SYSTEMS

SEWERAGE SYSTEM CONFIGURATION, FACILITIES AND OPERATION

Purpose of a Sewerage System

Sewerage systems collect and dispose of wastes (sewage) from living quarters, homes, apartments, industries, commercial establishments, and storms. Disposal is accomplished after treatment (usually) and discharge into a receiving body of water. The primary purpose is to protect public health and the environment.

DESCRIPTION OF A TYPICAL SEWERAGE SYSTEM

Sources of Waste

Sewage is discharged into the system by connections from many sources. What is called domestic sewage comes from living quarters and commercial establishments. Industrial waste is often treated or collected separately when its disposal into the domestic system may be dangerous or harmful. Storm run-off may enter the system, if it is a combined system, from catch basins, roof drains and drainage channels. All of this waste enters the system through pipe connections between the system and the facility where the sewage originates, commonly known as sewer connections. In some areas, storm run-off is collected in a separate system from the sewer system and is discharged into a receiving body of water.

The Collection System

This is the system of pipes that collects the sewage from the sources and conveys it to a central point for treatment and/or disposal. In contrast to the power and water distribution systems, the sewage collection system is not a network. Its configuration is in the from of a tree. The local portion of the system normally is designed to flow by gravity, i.e., the individual pipes are sloped downhill at a sufficient grade to carry the estimated flows and also to insure adequate scouring velocity. The grade or slope required depends on the size of the pipe. The pipe system generally follows the slope of the terrain.

Sewers that carry only domestic and commercial wastes are called "sanitary" sewers, those that carry only storm run-off are called "storm" sewers and those that carry both are called "combined" sewers.

Sewers are usually straight in both plan and elevation between manholes. The manholes, commonly spaced at about 300 feet along the sewer provide access to the sewer for maintenance and cleaning. Large sewers (5 feet diameter or more) can have manholes more widely spaced and horizontal curves and vertical grade changes are allowed. The local sewers flow into larger sewers sometimes called intercepter trunks or mains.

Sewage is often pumped, when the terrain dictates, through force mains routed around or over the obstacle. Inverted siphons are also often employed to pass under streams, canyons or other obstacles to maintenance of a uniform slope. Other variations from the conventional system include storage in or along the system, and overflow structures that allow discharge overland or to adjacent water courses when the collection system cannot handle the flow. Overflows are usually caused by storms which impose flows in excess of the capacity of the system.

Treatment

The treatment of sewage is a complex process, modern plants, though different, are as complex as any industrial or nuclear facility. They involve both controlled chemical and biological processes. The following steps in the treatment process are those most commonly found in modern treatment facilities and are described in the usual order the process train follows.

Bar Screens -- Removal of solid debris such as wood.

Grit Removal -- This step is necessary to protect equipment later in the process from abrasion and wear and to keep the grit out of sedimentation tanks and digesters. It is accomplished by controlling the velocity of the sewage such that the grit, which has a higher specific gravity than other solids, settles out but the lighter solids do not.

Grinding -- Provided in order to reduce the size of solids to manageable dimensions.

Separation of Solids -- Accomplished by settling (clarification) or screening, this step separates the solids that are not in solution and makes it possible to treat them separately. The settling or clarification process for removing or separating solids involves the introduction of sewage into a basin, either circular or rectangular, where it is held for a period of time sufficient for the solids to settle.

Biological Treatment -- This process utilizes the ability of organisms to break down the sewage into simpler compounds and reduces the demand for oxygen (Biochemical Oxygen Demand, BOD). It is accomplished in two ways as follows: aerobically (with oxygen) by processes such as activated sludge, trickling filter, and aeration; and anaerobically (without oxygen) in closed digesters in which the solids previously separated are continuously mixed in the presence of bacteria which do not require oxygen and produce gasses, principally methane and hydrogen sulfide. These solids are then removed by secondary clarification

Disinfection -- This step is intended to kill any remaining harmful bacteria and is accomplished by the use of chlorine, ozone, aeration and other processes.

Tertiary Treatment -- Occasionally secondary effluent is treated to a higher quality using coagulation, filtration, demineralization, or tertiary processes. This results in a reclaimed water with a quality nearly equal to potable water for use in industry, agriculture, and landscaping.

Disposal of Treated Solids and Effluent -- Liquids are normally discharged to a water course, to a groud water basin, or sometimes, if the treatment is incomplete, to a stabilization, evaporation, or infiltration pond. Digested solids (sludge) are dried on open beds or dewatered mechanically and deposited in land fills or sold as fertilizer or soil conditioner. The gas produced is either burned or utilized as fuel for heat or power.

ESSENTIAL AND NON-ESSENTIAL FUNCTIONS OF A SEWER SYSTEM

In emergencies such as might be caused by an earthquake or other natural disaster, some operations of sewerage systems may be discontinued temporarily. Some understanding of those operations that are essential will be helpful in post-earthquake investigations so that effort is concentrated on the important portions of the system.

The collection system must maintain flow unrestrained after the earthquake. Blockage of sewers can result in overland flow resulting in health hazards, disruption of traffic and sometimes flooding. If not quickly corrected, collection system failure may even call for shut down of the water system to prevent contamination. It should be noted that shutdown of the water system may not stop flow; ground water infiltration may continue flow in the pipe. Failure of pumps in a collection system will cause similar disruptions unless bypass or relief is available. Often, pump stations are constructed with limited emergency storage and bypasses to adjacent water courses. These may be allowed to function until repairs are possible, depending on the impact to the environment and health. Some stations have generators which can provide emergency power for limited periods. Post-earthquake investigators should examine these situations first, if possible, as they will be the first to be corrected and the evidence of the cause of the failure lost.

Treatment facilities, or specific unit processes can be shut down if they are inoperable. Complete shut-down could require a bypass of some kind or overland flow will result, with resulting health hazards and other complications. The time a treatment plant is allowed to be completely shut-down is critical, but it is during this time that observation by an investigator might be most productive.

Partial shut-down of treatment facilities is probably the most likely situation that an investigator may find. In such a situation, disinfection would usually be continued or placed back in operation as soon as possible and flow continuity maintained. Damage causing disruption of flow should be the first items to investigate, as they will be the first to be corrected. Functions which might be temporarily shut-down include: grit separation, grinding, settling of solids, biological treatment (may require several weeks to restart process), sludge transfer.

DANGERS IN AN EARTHQUAKE DAMAGED SEWERAGE SYSTEM

Collection System

Blockage of flow or pump failure can be serious, should normally be readily apparent from the ground surface, and will cause overflow from manholes, and backup into living quarters or commercial and industrial facilities.

Gas in sewers is a serious hazard and extremely dangerous. A blocked sanitary sewer will become a closed digester and begin to generate methane gas in only a few hours. Methane is odorless and highly explosive in sufficient concentration. Whole blocks of sewer have been known to explode. Damaged sewers should never be entered until they have been opened, thoroughly ventilated and checked for gas concentration. Even with positive gas detection results, underground chambers should not be entered without proper safety precautions, such as additional personnel, lifting devices, etc. Hydrogen sulfide is also generated in a closed sewer where the oxygen has been depleted. Hydrogen sulfide has the well known "rotten egg" odor, but, unfortunately, the human nose adjusts to the odor, and after a period the nose no longer detects it. The result can be fatal. Pump stations are typically underground and are therefore subject to similar dangers as sewers. In addition, mechanical equipment such as drive shafts, pulleys, belts, etc. can become loose or misaligned and are subject to failure.

Treatment Facilities

Methane, hydrogen sulfide, and chlorine gas can be prevalent in treatment facilities that have been damaged. This can be the result of equipment failure or mechanical or electrical breakdown. Gas control facilities, gas storage tanks, engines, waste burners, etc. can be damaged by an earthquake resulting in release of gasses sufficient to cause critical concentrations. With the presence of electrical power and wiring the possibility of fire and explosion is always possible. Most parts of sewage treatment plants are classified as hazardous by the National Electric Code.

Flow stoppage in a treatment plant resulting in overflow to the surrounding land can be hazardous from a health standpoint, but also cause washouts, slides and erosion. In addition, many treatment plants use chlorine and other chemicals which can be a hazard to investigators. Check with operations personnel to determine the hazard potential.

SEWERAGE SYSTEM EQUIPMENT AND THEIR FUNCTIONS

Collection System Equipment

Pumps

Pumps are used to lift raw sewage when gravity flow is not possible or practical. Pump types used are horizontal and vertical configurations of non-clog, mixed-flow, and other similar proprietary variations of the open impeller design. They are driven, usually, by electric motors and often provided with gas or diesel standby drives or generators. Motor starters are across the line or reduced voltage, depending on the size, and can be manually or remotely operated. Pumps are controlled by float or probe actuated level switches, often as part of a computerized plant control system.

Gates and Regulators for Flow Control

Overflow and bypass are often controlled by shear gates, valves (plug, ball, gate, etc.) or other devices such as vortex flow controllers. This equipment is used when the capacity of the collection system is expected to be exceeded at certain times and under certain circumstances. The purpose of this type of control is to prevent overloading of sewers and treatment facilities and control discharge to storage facilities or occasionally directly to receiving waters. Such gates and valves are operated electrically, manually, hydraulically, or pneumatically. They may be stopped and started by manual switches, but usually are controlled remotely by telemetered signal or computer.

Treatment Plant Equipment

Grit Removal Equipment

Grit is separated from the plant influent by a variety of equipment. Earlier plants used channels designed to provide the proper velocity through a range of flows and washed the settled grit through a cascade system or a grit washer. More recent developments for grit removal include aerated grit chambers and vortex separators and washers

Grinders and Bar Screens

Older plants used comminutor type grinders which consist of a slotted cylinder through which the sewage flows while rotating teeth engage the slots. More recent developments include bar screens which can be manually cleaned, but are usually mechanically cleaned, where the screenings are passed through shredders or grinders.

Screens

Fine solids are often removed by screens, usually of the rotary drum type. Some screens are set on a slope and the sewage enters the upper end on the inside. The solids remain on the inside of the drum and gradually work to the downstream end where they drop into a hopper and are pressed to remove excess liquid. Other screens apply sewage to the outside of the drum and allow liquid to fall through. Micro screen or fine screens are occasionally used, but normally only for special applications.

Clarifier Mechanisms

These mechanisms consist of influent columns or ports, scrapers to move the settled solids to a hopper, skimming mechanisms to collect or concentrate the floating materials and overflow weirs or collectors for discharge of the clarified effluent. Scraper or skimmer flights propelled slowly by chains or rake arms, unless specifically designed for the type of forces an earthquake might impose, are vulnerable to shaking and lateral forces of earthquakes.

Sludge Pumps

These pumps are specifically developed to pump liquids with high concentrations of solids. They may be positive displacement or open impeller types or variations thereof. They are normally driven by electric motors and are usually manually controlled. The pumps are normally located in pits adjacent to sludge hoppers in clarifiers or digesters and are used to transfer sludge.

Mixers

Mechanical mixers are often installed in digesters and other solid process tanks to keep solids in suspension and promote uniform digestion. They are also sometimes found in liquid process basins where it is desired to prevent solids from settling or segregating. These mixers are turbine impellers with straight, curved, pitched or vaned blades mounted on a shaft which is turned by motors generally mounted above the material to be mixed. Horizontal side mount (through tank wall) mixers are also occasionally used as are horizontal submerged in-tank mixers.

Aeration Equipment

Aeration is a common process used in sewage treatment to provide oxygen for aerobic biologic decomposition. Air is introduced through diffusers consisting of porous tubes or nozzles of various types. This requires blowers of relatively large capacity which may be centrifugal or positive displacement, usually electric motor driven and manually controlled for the most part since most aeration is a continuous process. Aeration may also be accomplished mechanically with surface aerators or submerged turbines similar to those previously described for mixers.

Thickening

This step in the treatment process, intended to increase the concentration of solids prior to other processes uses gravity belt thickeners, centrifuges and/or dissolved air flotation. The latter requires a basin provided with skimmers, sludge scrapers and air diffusers or nozzles.

Dewatering Equipment

Sludge is dewatered, usually after digestion, by the use of belt filter presses, centrifuges, vacuum filtration or beds, gravity beds, or plate and frame presses. There are many types and variations of this equipment. It is usually manufactured as self contained units, but may be subject to damage if not properly anchored or earthquake qualified.

Digester Equipment

Anaerobic sludge digesters are closed tanks with many types of equipment associated with their operation. Many plants will include at least one tank with a floating cover which may be vulnerable to earthquake damage. Anaerobic digesters produce gas, so the associated gas control and utilization equipment will usually include gas dryers to reduce moisture in the gas, gas control and metering equipment, gas burners for waste gas and gas-driven engines to provide power and save energy. The gas piping associated with this equipment may also be vulnerable to damage. Aerobic digesters use aeration equipment to further digest the sludge. This process usually does produce hazardous gases. Sludge is dewatered using methods similar to that of anaerobically digested sludge.

Disinfection Equipment

Chlorine gas is generally delivered to plants in 150 pound cylinders, tank containers or tank cars. The cylinders themselves may be subject to damage if not properly anchored. Since chlorine is poisonous even at low concentrations, extreme care must be taken in investigating such facilities after an earthquake. All chlorination facilities will or should be equipped with appropriate masks and safety equipment. Chlorine is fed through machines which dissolve the gas in water and this solution is applied to the sewage. The piping between tanks, chlorinators and the point of application is vulnerable to damage. Some plants use sodium hypochlorite or calcium hypochlorite in place of liquid-gas for safety reasons or availability. These compounds are dissolved in water in mixing tanks and applied to the sewage by the same procedure as the gas water mixture.

Ozone has become an often used alternate to chlorine for disinfection in recent years because of the carcinogens resulting from the reaction of chlorine with some chemicals and organics in water or sewage. Ozone is usually generated on site. It can be applied to sewage as a gas requiring facilities to insure proper contact of the ozone with the sewage. The generating equipment requires power, as the ozone is generated by passing air through electrodes. The generators are generally self contained units. Also, ozone can be delivered by rail tank car or tank truck to the treatment plant site.

Controls

Sewage treatment is controlled in many ways, from completely manual operation to, in very recent plants, almost complete computer control. The information on which the control is based is obtained from a myriad of measuring devices and meters including probes, floats, flow meters, chemical and solids concentration measuring devices, etc. This information is assembled or brought together in a control room where it is monitored and operating decisions are made by the operators or computers. The appropriate manual or remote switching of valves, gates and other equipment are directed from this control center. The control room would be the first place to visit in beginning the post-earthquake investigation of a treatment plant.

SOURCES OF SEWERAGE SYSTEM DAMAGE INFORMATION

Geologic and Soils Information on the Region and Site

This type of information on a specific site may be difficult to obtain. General information on a region should be available through USGS maps and the nearest USGS office. Local geologists and soils engineers would also be a good source. Local soil information can be obtained from well drillers and water and sewer construction contractors. The latter may be difficult to find immediately after an earthquake, as they will be the first called on to make repairs or drill test holes and are probably best found by keeping an eye out for their equipment.

Visual Observation Above Ground

Above ground evidence of sewerage system damage might include indications of surface flooding, damage to above ground structures such as pump housing, curbs, sidewalks, or pavement, and evidence of subsidence, liquefaction, slides, etc. Such surface indications may not mean damage to sewers, but will point to the first places to investigate. Also, lack of surface water does not mean that sewers are undamaged.

System Operating Personnel

Obviously, the system maintenance superintendent or manager would be the first person to contact, but such persons may be extremely busy and preoccupied. If this is the case, the supervisor and workers in the field can be a good source of damage information. So also can the person in charge of the equipment and supply yard who would know where repairs were being made and probably what type of damage is being repaired. Examination of flow charts and records, including power consumption records, before, during, and after the earthquake may divulge clues as to what happened and even where the damage occurred. A visit to the maintenance and operation headquarters should be made at an early stage.

Local Newspapers

Careful reading of articles about the earthquake may indicate the location and area where most damage occurred, where problems with sewer house connections have been reported. Reports on discussions with officials and workers may also provide damage information.

SEISMIC PERFORMANCE OF SEWERAGE SYSTEM FACILITIES AND EQUIPMENT

Underground Facilities

Piping

Vitrified clay and concrete sewer pipe, at least in the smaller (under 36-inch) sizes is manufactured in short lengths (3 to 6 feet). Asbestos concrete and PVC pipe is usually installed in 10 to 12 foot lengths. Bell and spigot joints caulked with cement mortar were common until about 30 years ago when the mechanical compression joint using an O-ring or compression ring of polyvinolchloride or polyurethane came into common use. This joint provides some flexibility and practically all sewer pipe is now installed with this type of joint. Large sewers may be cast-in-place concrete. Brick sewers are found in older cities. Some large sewers are pile supported. Ground deformations caused by lateral spreading, uplift, liquefaction, movement at a fault crossing or slides generally are the cause of sewer damage in an earthquake. Ground shaking and wave propagation effects can also cause damage. Where sewer pipe enters or joins a massive structure, earthquake damage can be expected, unless special precautions are taken in design and construction. Intense or prolonged ground shaking can cause pipe to crack or collapse and joints to open or compress resulting in infiltration, exfiltration, or blockage.

Not many earthquakes in this country have occurred where extensive sewerage systems were in place. The best example is the 1971 San Fernando earthquake. The sewerage system in the Knollwood-Sylmar-San Fernando area was extensively damage. This area lies on the alluvial or plain at the foot of the San Gabriel and Santa Susana Mountains. Extensive faulting and uplift occurred in the area. Of the approximately 110 miles of sewers in this area, roughly 25 miles were so badly damaged that reconstruction was required. Damage consisted of broken pipe and joints, pulled joints, and changes in grade. Pipe with flexible joints performed much better than rigid (cement mortar) jointed pipe. In the 1989 Loma Prieta earthquake, extensive sewer damage occurred in the Marina District of San Francisco and in Santa Cruz, primarily in areas where there was differential settlement and/or liquefaction.

Manholes

Most manholes are precast concrete pipe sections installed vertically. Older installations may be brick. Damage to manholes will usually be limited, although extensive faulting or uplift can cause failure. Manhole walls sometimes are cracked and rings and covers shifted out of alignment. The junction between the sewer pipe and the manhole is the critical point, probably due to the difference in mass. Over 100 manholes required repair after the 1971 San Fernando earthquake.

Pump Stations

Pump stations housed above ground, when built to modern seismic standards perform well. Unreinforced masonry buildings may crack or collapse. Suction and discharge pipe connections to the building may fail if flexible couplings are not provided. Wet and dry wells, particularly if deep underground, may crack under heavy shaking and may become misaligned if subsidence or liquefaction occurs. The sliding of unanchored equipment and differential settlement can damage piping.

Treatment Facilities

Basins and Vaults

Walls of these structures may be subject to unusual or unanticipated lateral forces if they are buried. Connections between structures of different mass may be subject to damage such as the point where a channel joins a basin. Expansion or contraction joints in these structures, since they are designed to allow movement, may crack or spall due to differential displacement if the movement is greater than the designer anticipated. Basin and vault covers perform similarly to building roofs during earthquakes, except that sloshing of the contained liquid can damage or lift covers off their supports. Floating digester covers were particularly vulnerable in The Loma Prieta earthquake. Sloshing may damage baffles in clarifiers. Refer to the sub-section on tanks in Section 12 for further information on tanks. Long galleries are subject to cracking from the earthquake forces.

Equipment

Inadequate anchorage is the most common cause of equipment failure during earthquakes. Chlorine tanks, chemical tanks, presses, pumps, valves, motors, control cabinets, batteries, etc. are all vulnerable when not properly anchored. Mixers, scrapers, clarifier mechanisms and similar units are subject to damage or displacement, particularly the moving parts such as scraper arms, tracks, chain belts, long shafts and moving bridges. Baffles falling on sludge scrapers have jammed them causing failure. Earthquake qualification of equipment has recently been introduced as a means of assuring that the installed equipment has been designed and tested to withstand earthquake forces.

GUIDE FOR INVESTIGATING SPECIFIC FACILITIES AND EQUIPMENT

Examination of Underground Piping -- CAUTION-CHECK FOR GAS

System Configuration and Damage Overview

The first step in the post-earthquake investigation of the sewers in a collection system is to obtain a plan map of the system showing sizes and locations of all pipe and manholes. Information on where the damage has occurred obtained from the system personnel (see earlier section entitled "Sources of Sewerage System Damage Information") should be added to this map. It is doubtful that an outside investigator, for whom this guide is prepared, will be the first one on the scene of an earthquake. Therefore it is probable that operating and maintenance personnel will already have determined what has been damaged and what type and where ground displacement, faulting, subsidence or uplift has occurred. This information should be also added to the map. From this data, the area within which the investigation should be concentrated is established. A standard form for recording information from the field investigation should also be developed in this preliminary or organizational step.

Damage to sewer pipe may not come to light for some time after the earthquake when roots have grown into cracks, leakage shows on the surface, flow discontinuities develop and other evidence accumulates as pipe is excavated and repaired. Therefore, it is very important to establish contacts so that additional information can be obtained when it becomes available, possibly several months after the earthquake. The inspection procedure described below is required for a complete examination of a collection system. The investigating team for which this training guide has been prepared will probably not have the time or manpower to perform the inspections described but can develop preliminary estimates of damage and encourage the system operating personnel to go through all the steps and must arrange to obtain the results later as the operating personnel develop it. In particular is it important to stress the need for complete and detailed records so that statistics can later be developed regarding the number and frequency (failures per mile for example) of various types of breaks. Detailed damage reports will have to be prepared to get reimbursement from FEMA. These will often be based on TV inspection of the system. Attempt to find a source for FEMA data.

Manhole Examination and Sewer Lamping

The first field step in investigating the sewage collection system is a field examination of the manholes. Manhole lids should be opened, any apparent structural damage noted, vertical alignment checked and sewage flow noted. Beside location, detailed information should be recorded and include at least the following:

WARNING: Do not enter any sewer or vault without first checking for toxic gas.

Diameter of ring and cover
Size--usually diameter at base
Depth--edge of rim to flow line
Type--best described by sketch
Material--brick, concrete block, precast concrete, cast in place concrete
Inlet pipe size
Outlet pipe size
Laterals or drops--size and location
Nature of flow--free flowing or stagnant
Depth of flow--above flow line
Plumbness -- estimated out of plumb in inches per foot in direction of main sewer and at right angles
Structural damage, cracks or breaks
Signs of settlement or movement of entire manhole

If damage or unusual sewage flow or lack thereof is noted, during the manhole inspection, the sewer should be "lamped". This procedure involves opening two adjacent manholes and sighting down the sewer between them. This should not be done by the investigator unless proper safety procedures are implemented. Sometimes lights and mirrors are used in this procedure. A full lamp usually indicates no damage, whereas a partial or no lamp probably indicates damage. Where damage is indicated, the approximate location should be estimated and recorded and surface evidence at the estimated location looked for. In a system of any size, operation and maintenance crews may already be engaged in this procedure when the investigator arrives. At this stage system personnel are primarily interested in getting the system back in operation, so detailed records and locations may not be kept. In this activity, the investigator should make sure the records are kept accurately and completely, even if the investigator must do it personally.

Once damage has been located, the sewer should be rodded. Maintenance personnel should have rodding equipment, power driven or manual. Such work is probably beyond the ability of the investigating team. If rodding does not correct the damage, the sewer must be uncovered to determine the damage, or the sewer can be examined with a video camera with which many sewer agencies are now equipped. It is important that detailed information for each failure discovered include, beside location data, at least the following:

Pipe size
Pipe material
Type of joint
Depth of pipe
Type of soil
Description of break or damage
Age of installation
Preliminary evaluation of cause of damage
Relation of faults, uplift or subsidence zones, slides, etc.
Other information of interest in analyzing the failure.

INVESTIGATION OF TREATMENT PLANTS AND PUMP STATIONS

Preparation for the Investigation

This work should be approached as outlined in Section 4 under the subheadings "Documenting Damage", "Failure Modes" and "Factors Contributing to Equipment Failure". As in preparing for the investigation of a sewage collection system, discussion with operating personnel to determine what equipment has failed or is shut down is the place to start. A history of the failure or shut down and of the equipment itself is desirable. Reasons for the shut down, or opinions as to the cause will be helpful. From this preliminary information, a list of significant equipment failures to be examined in the field can be compiled.

FIELD INVESTIGATION OF FAILED EQUIPMENT

Information Needed About Failed Equipment

The first step in the field investigation of a particular piece of equipment is to verify the information previously obtained regarding make, model, size, capacity, etc. Next, the reasons for the shut down should be ascertained as accurately as possible. The information obtained from operating personnel may not be correct, as it may have been second hand or without full examination. It is important to look for secondary causes of failure such as lack of power, piping damage, control system failure, fuel line damage, lubricating systems, etc. Equipment that was not damaged may be shut down because its continued operation will cause difficulties in other parts of the system, i.e., a pump might be shut down because the discharge line is broken and would cause flooding, not because the pump has failed. Determine if the equipment was properly anchored, and if not how did the anchorage fail.

Determining the Cause

Pictures and detailed descriptions will be of great help later, when the investigator is trying to piece together data from many pieces of equipment under the pressures of time and the confusion accompanying the disaster. Before leaving the site of any equipment failure, the investigator should make a preliminary determination of the mode of failure and its cause based on the information he has at the time. Few investigators, if any, will fully understand all the types of equipment that might be examined in a sewerage system, and assistance should be sought from manufacturers, maintenance personnel and anyone else who might be able to contribute.

INVESTIGATING TREATMENT PLANT STRUCTURES

This type of investigation requires investigation of the basins, tanks, vaults, flumes, conduits, and yard piping in the plant. Guidance as to what to examine should be available from the plant operators, some of whom might have been present during the earthquake. Look for joints and cracks in concrete. Check the alignment of walls, channels, etc. and look for differential movement between structures. Investigate the pipe and conduit connections to structures. Study pipe hangar performance. Attempt to sketch, photograph, or describe the movements causing the damage and determine the cause, if possible. Look for surface water or typical sewer odors.

CASE STUDY OF AN EARTHQUAKE INVESTIGATION OF A COLLECTION SYSTEM

Description

The investigation selected for this example was conducted by the Subcommittee On Water and Sewerage Systems of the NOAA/EERI Earthquake Investigation Committee. The investigation is described in a report entitled "Earthquake Damage to Water and Sewerage Facilities" and was published in 1973 as a part of Vol. II of the report on the San Fernando Earthquake of Feb. 9, 1971 The investigation covered the sewers in the Knollwood/Sylmar/San Fernando area shown in Fig. 8. 1. The area included 110 miles of sewers.

Damage Survey Procedures

The basic procedure followed in the damage survey was to first select the area to be studied in detail on the basis of surface indications and operation reports of damage. This area was then surveyed in detail by first uncovering each manhole to determine damage and if sewage flow existed. If damage was observed, the manhole was entered and the sewer inspected with lights. Sewers indicating damage were then rodded by hand and power rodding equipment to clear stopped lines and make a better determination of the extent of damage. This procedure indicated those lines that were completely collapsed or contained broken pieces. However, cracked sections, damaged joints or other damage that did not impede the passage of the rodding equipment could not be detected by this method. A more sophisticated means of inspection was called for, and with the participation of the Corps of Engineers, all 110 miles of sewer in the selected area were examined by video camera, requiring up to 11 television crews.

The investigation indicated that the types of sewer damage that occurred included broken pipe, broken joints, pulled joints, changes in grade, and shifting and cracking of manhole structures. Over twenty percent of the sewers in the area of the study were considered so badly damaged as to require reconstruction.

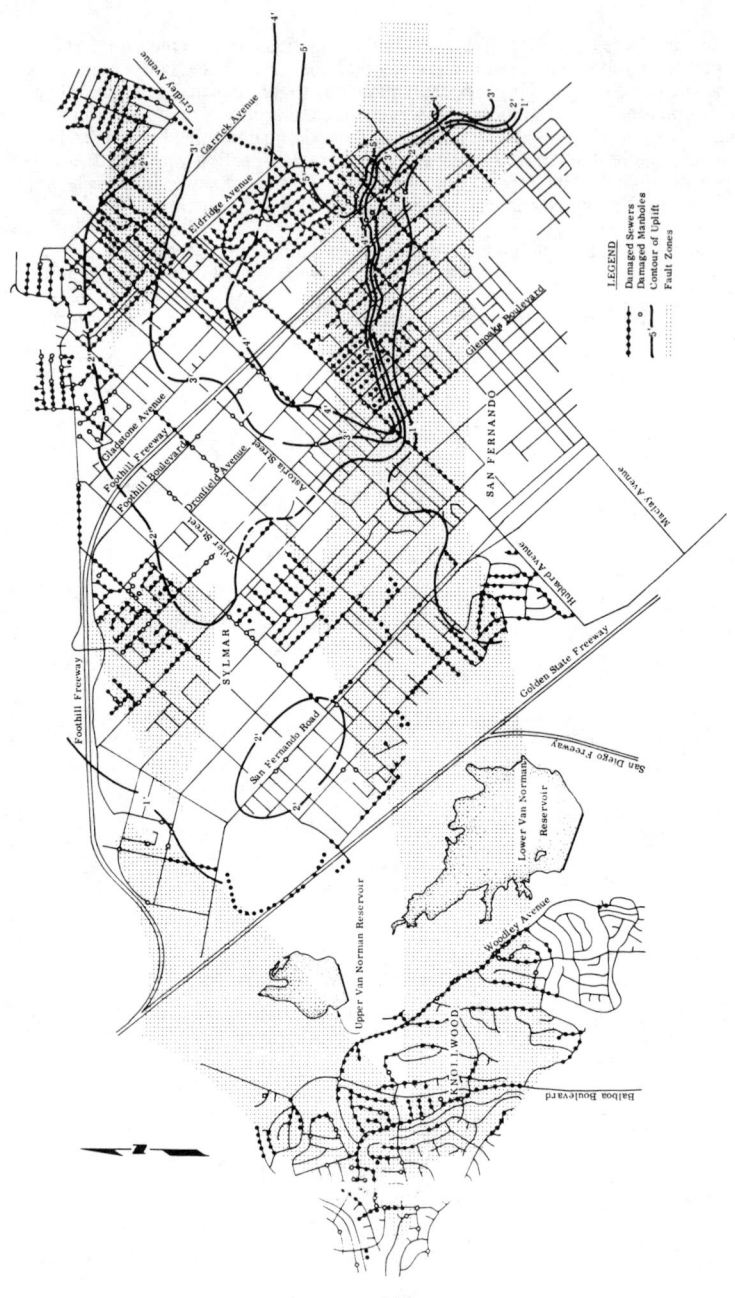

Fig. 8.1 Map showing major sewer damage in San Fernando after the 1971 earthquake.

SEWERAGE SYSTEM CHECK LIST

Key information is indicated with an *

Manholes
    Diameter of ring and cover
    Size--usually diameter near base
    Depth--edge of rim to flow line
    Type--best described by sketch
* Material--brick, concrete block, precast concrete, cast in place concrete
    Inlet pipe size
    Outlet pipe size
    Laterals or drops--size and location
    Nature of flow--free flowing or stagnant
    Depth of flow--above flow line
    Plumbness--estimated out of plumb in inches per foot in direction of main sewer and at right angles
* Structural damage, cracks or breaks
* Signs of settlement or movement in the ground adjacent to manhole
    Age or date of installation

Sewers
* Pipe size
    Pipe wall thickness--this may be difficult to see directly and may need to be obtained from operating personnel
* Pipe material--common materials are vitrified clay, concrete, asbestos cement, PVC, cast iron, ductile iron, concrete cylinder, and steel
* Type of joint--joint types in common use are bell and spigot with cement mortar, bell and spigot with rubber ring, tongue and groove with cement mortar, tongue and groove with rubber ring, push-on and many proprietary types
* Age of pipe or date of installation
    Depth of burial
    Type of soil--general non-technical soil types might be
        Soft soil
            Loose sand
            Unconsolidated silt
            Loam
            Mud
            Dump fill
        Firm soil
            Gravel
            Consolidated sand
            Consolidated silt
        Rock
    Type of bedding material--if special bedding was provided (this information will probably have to be obtained from the operating or maintenance personnel or from the specifications for the original construction
    Type of backfill--some times special backfill material is used up to the spring line or a few inches over the top of the pipe and this information will also have to come from operating or maintenance personnel or the original plans and specifications
* Ground water elevation
    Drainage--surface or underground

* Description of break or damage
* Preliminary evaluation of cause of damage
* Surface indications of ground movement including extent of
    Faults
    Uplift
    Subsidence
    Slides
    Lateral spreading
    Soil cracks--size and extent
  Did damage disrupt or block flow
    Other information of interest in analyzing the failure.

Treatment Plants and Pump Stations
    Investigating equipment failures
        Interview operators for failed equipment
*   Determine or verify for failed equipment - general description
    Make
    Model
    Size
    Capacity
* Verify reasons for shutdown
* Look for secondary causes of failure
    Lack of power
    Piping damage
    Control system failure
    Fuel line damage
    Lubricating system failure
    Other causes
* Review Anchorage (see Anchorage check list)
    Attempt to determine cause of failure
    Seek assistance from manufacturer, maintenance personnel, designers, consultants.

    Investigating Treatment Plant or Pump Station Structures
* Interview operator and maintenance personnel
* Examine joints
* Survey for cracks and broken supports
  Check alignment of walls, and channels
* Look for differential movement between structures
* Investigate pipe and conduit connections to structures
* Evaluate pipe hangar performance
* Sketch or describe movements causing damage
  Make a preliminary determination of the cause of the failures
* Check for tank damage, damage to baffles and digester covers
  Date of construction
  Were seismic design codes used in its construction?

9. Transportation Systems

   Airports
      System Configuration
      Major Operating Units and Facilities - Airfield-side
      Major Operating Units and Facilities - Land-Side
      General Facilities
      Earthquake Performance of Airport Facilities
      Tips On Conducting Post-Earthquake Investigations
   Ports
      Port Description
      Port Facilities
      Typical Earthquake Damage Experienced By Port Facilities
      What to Look For
      Case Study
   Highways
      Highway Bridges
      Tunnels
      Roadways
   References
   Airport Check Lists
   Port Check Lists
   Highway Check List

9. TRANSPORTATION - AIRPORTS

SYSTEM CONFIGURATION

The airport which is described here represents a large hypothetical international airport so that many of the facilities described here may not be present at smaller airports or even some large airports. Functions at some airports may be combined. It should be noted that airports are usually divided into airfield-side and land-side operations by airport personnel and this approach will be used here. The size and complexity of airport operations is not generally appreciated. A large airport will have many of the function of a small city, and the total number of employees located at the airport approaches that of a small city. For example, Los Angeles International Airport has over 53,000 personnel working for the various organizations on the airport.

MAJOR OPERATING UNITS AND FACILITIES - AIRFIELD-SIDE

Control Tower Air Traffic Control Facilities

The air traffic controllers located in the control tower and most of the active air traffic control equipment, such as radar , radios for communication with aircraft, and radio beacons are under the jurisdiction of the Federal Aviation Administration (FAA). The monitoring and control (M&C) of aircraft can be divided into three or four areas. The country is divided into 23 regions, each with a air traffic control center that monitors aircraft in air traffic corridors for long distance flights. The airspace around major metropolitan areas is controlled by an air traffic control center. As an aircraft approaches an airport, M&C are passed to the control tower at the airport. Once an aircraft is on the ground, its M&C is passed to airport ground control. Each of these control functions have designated radio frequencies and alternated frequencies over which controllers and pilots communicate. Commercial aircraft are fitted with transponders which allow controllers to identify the aircraft and indicate its location and altitude, and speed. This information is used by FAA computers to warn of collision courses and is displayed on radar screens used by controllers.

Radars that provide information for airport operation are typically located at the airport; however, if there is more than one major airport serving an area, the radar may be located away from the airport. There are special radars that provides detailed information about the position of the aircraft on its final approach. Additional ground based radio antenna provide signals to the aircraft that indicate the position of the aircraft along its glide path. Under normal flight operations, visual observation by the air traffic controllers is not normally used. In the event of the loss of electronic support, landing operations could be done using visual control, but aircraft separation and over all efficiency would be significantly impaired.

Runway approach and landing lights are under FAA jurisdiction but support structures (for approach lights off shore at the ends of runways adjacent too water) may be part of the airport's facilities.

Emergency power for the control tower and air traffic control facilities is operated by the FAA. Smaller airports may use emergency power provided by the airport.

Ownership and the physical location of telephone communications equipment for the tower will vary from airport to airport.

Control towers are usually surrounded by windows to provide controllers maximum visibility. Special attention id given to reduce outside noise in the control tower. Because control towers are typically tall structures, they have an elevator. Critical equipment would include radios and associated antenna for communicating to aircraft, radar and associated computers and radar display screens, and communication links to other FAA centers that they may have to contact.

Many airports may have two control towers, a new one that repaces the old. The older tower may still be FAA certified and used for general aviation or as a backup.

Runways and Taxiways

Runways are usually reinforced concrete with asphalt cover and must be able to carry the heavy load imposed by modern large aircraft. Runways typically are 8,000 to 11,000 feet long. Larger airports often have two sets of parallel runways.

Fire Response Units

Special fire response units are used in the event of an airplane crash, fuel spills, and fires in other facilities at the airport.

Airport Security

Airfieldside and landside security may be run out of one office but special consideration of baggage and cargo and passenger is needed. If the airport is a port of entry, there will also be Federal Customs and Immigration authorities.

Cargo Facilities

Cargo facilities would include special equipment to unload cargo from aircraft, temporary storage facilities, freight loading facilities, and freight forwarding agents.

Passenger Luggage Facilities

In addition the equipment to facilitate the loading and undloading the unloading of aircraft and the means for transporting baggage on the airfield, there are extensive automated baggage handling facilities in the terminals.

Fuel Storage and Supply

In addition to the storage of large quantities of jet fuel, there are pipelines to deliver the fuel to the aircraft. Fuel is also needed for the many airport emergency and service vehicles. In addition to storage, there is a need for emergency power for these facilities.

Airline Food Services

Aircraft food service companies provide thousands of meals a day that are served on aircraft. These organizations will have extensive food preparation areas and food storage facilities.

Aircraft Service Facilities

In addition to maintenance facilities for the repair for aircraft, service crews are needed for normal aircraft turn-around service.

MAJOR OPERATING UNITS AND FACILITIES - LANDSIDE

Passengers

While not an operating unit of the airport, there may be thousands of passengers at the airport at any time. In the case of an emergency they are at risk of injury and due to disruption of transportation to and from the airport, they may have to be cared for.

Terminal Facilities

The terminals not only provide for most of the passenger needs while they are on the ground, they are the interface between the airfield and land-side operations. They provide the jetways for loading aircraft, and roadways for dropping off and picking passengers up. The terminals also provide centers for operations for many of the airfield-side and land-side functions.

Police

Police perform their normal functions for landside operations, although traffic control may take most of their time.

Road Systems

The airport will be provided with an extensive road system to transport passengers to and from the airport. Because of the complexity of the system, there will typically be many elevated roadways and overpasses.

Railroad Systems

More airports are being provided with light rail systems for the transportation of passengers. These may be elevated or subway systems.

GENERAL FACILITIES

Non-Aircraft Communications

An airport has extensive communication facilities other than those needed to communication with the aircraft. The airport will probably have a communications center. Some of the functions that they will be responsible for are for all telephone communications at the airport. These may be Centrex system where the equipment is physically located at the central office of the telephone company, or a large private branch exchange (PBX) with a large switch located at the airport site. The airport may have its own Public Safety Answering Point (911) center that answers 911 calls made at the airport and responds to them with airport resources. There may be a paging system for contacting service personnel. There will be several radio networks that may have repeaters located around the airport. These networks would deal with fire, security, and service. Additional communication links are maintained with area emergency response services such as police and fire mutual aid, Office of Emergency Services, etc. Communications systems are usually supplied with emergency power.

There is also a need to be able to communicate with passengers at the airport following and earthquake. There is a need for intercom that works after the earthquake and a means for communicating with the passengers when they are outside of the terminal in the event that it has to be evacuated.

Water and Sewage Systems

Airports will have extensive water and sewage systems. They may also have water storage systems for fire suppression. The terminal will have a sprinkler fire suppression system.

Power Systems

Some airport may be provided with power from more that one substation as a means of enhancing reliability. In addition, there are many emergency generators. For example, San Francisco International Airport has 14 diesel generators for emergency power. Other organization will also have their own emergency generators. For example, the FAA, fuel supply company have their own units.

Airport Services

There are several different types of services provided at an airport. There will be general maintenance, grounds maintenance, cleaning, food services, gift shops, etc.

EARTHQUAKE PERFORMANCE OF AIRPORT FACILITIES

Control Towers

In Anchorage in the 1964 Alaskan earthquake the control tower completely collapsed.

Control tower windows have been one of the most vulnerable parts of the airport and have been damaged in many California earthquakes and their aftershocks. Because of the mild weather in California, the tower could still operate but the noise level in the tower because of the missing windows had significantly disrupted operations as it interfered with communications. At San Francisco International Airport after the Loma Prieta earthquake, temporary windows had to be installed before normal operations could be started due to noise. The desirability of a unobstructed view from the tower leads to slender roof support columns that are flexible. This flexibility can cause the loss of windows. Windows have been damage at 2 airports in the Loma Prieta earthquake, and at the Whittier Narrows earthquake and aftershock.

In addition, equipment is often not anchored in the tower so that radar monitors can fall to the floor and be damaged. Suspended ceilings and sound proofing above the ceiling frequently fall, disrupting operations and injuring controllers.

The elevator that services the control tower is often damaged by the earthquake. The loss of water may put lavatory facilities in the tower out of service. This coupled with the loss of elevators may be more than an inconvenience for a tall tower.

The tower and the good view of the airport is often used for control of aircraft on the ground.

Many airports have two towers, an older tower that may be used for general aviation or for certain types of flight operations and is FAA certified, and the new tower that replaced the old one. In an emergency the old tower may be used, although it may not have the most advanced support equipment.

Emergency Power

Emergency power is often poorly installed so that the units are on vibration isolators without snubbers, the fuel system is unanchored, the fuel system may require a fuel pump that is not emergency powered, the fuel that is stored had gone bad, emergency batteries to start the emergency generator are unanchored, not all facilities that need emergency power have it. This has been the case for radio repeaters and fuel supply for aircraft and emergency vehicles. At the Burbank Airport, the PBX telephone system had no emergency power so that all telephone and the paging system, which used the telephone system, did not operate.

Dry-type transformers used for regular power and emergency power were damaged when the unanchored core in the transformer enclosure moved.

Runways and Taxiways

There has been extensive severe damage to runways and taxiways due to liquefaction. This is typically at airport constructed on fill adjacent to a body of water. Runways and or taxiways were damaged at SFIA, Oakland International Airport and Alameda Naval Air Station in the Loma Prieta earthquake. Runway approach light support structures that extend in to San Francisco Bay were damaged at SFIA in the Loma Prieta earthquake.

TIPS ON CONDUCTING POST-EARTHQUAKE INVESTIGATIONS

Its important to see several key people at an airport. During the emergency response period immediately after the earthquake, key airport personnel will probably be too busy to talk, but observation of operations and talks with airport operating personnel can be useful. In talking with various operating personnel at the facility, focus on their activities and those things that are effecting their activity, rather than an overview of problems.

While the airport manager should be visited, it is suggested that you see this person last as if permission is denied to visit the facility it would be improper to see other. A good starting place is chief electrical maintenance. He will know about power problems and may know about communications problems. Chief of mechanical maintenance would know about water and sewage problems and may know about structural problems. Others would include chief of communications, an FAA control tower representative, chief civil engineer, head of operation, airport manager.

In addition to damage, try to get measure of impact on operations, flight per hour, etc. Also check if there were any operation related to the earthquake, such as relief, medical evacuations, bringing in of technical support, impact of media and VIP visits, etc. Also get key statistics on the facility such as number of runways and length, normal number of operation , a measure of the freight passing though the airport and the number of passengers. Ideally it would be desirable to get this data on a weekly base at that time of year rather than figures for the entire year. Peak capacities would also be useful. If there is damage distributed around the facility, try to get a map, possible literature on airport development, so that damage sites can be located.

9. TRANSPORTATION - PORTS

PORT DESCRIPTION

A port (or harbor) is a facility where ships can take on or discharge cargo. Commercial harbors are an important link in the transportation portion of a region's lifeline because they provide a cost-effective method for moving large quantities of cargo into and out of a region. In many cases shipping goods through a harbor is the only means by which raw materials or manufactured products can be transported into or out of a region. As an example, current energy demands mandate that the United States import oil from foreign sources. This imported oil is transported exclusively by ships, and the oil must be transferred through harbor facilities at various U.S. ports.

Commercial harbors associated with the transportation of cargo and people have five primary engineered components. The first component is the navigation channel that has often been dredged to obtain proper depth. The second component is the pier or wharf (sometimes referred to as a terminal or berth) where ships dock and freight or passengers facilities are actually transferred to or from the ship. This portion of the harbor has high capital expenditure including specialized equipment to expedite the cargo transfer process. Local storage facilities are the third component of a harbor. These facilities are located on or adjacent to the pier and they provide temporary storage for incoming and outgoing cargo. The fourth component of a harbor is the transportation systems, roadways, railroads, and pipelines, that link the harbor with inland transportation systems. Finally, the harbor must interface with other lifelines, such as power, communication, water, and sewage systems. This interface forms the fifth component of a harbor facility. Note that damage to any of these components, if severe enough, can render the harbor inoperable for transportation.

Other activities that can be found at harbor facilities include the processing of raw goods such as fish; shipbuilding and repair; and preparation of goods, such as scrap metal, for shipment. Harbors often have tourist attractions (restaurants and shops) along with marinas for pleasure craft. Although the tourist attractions and marinas are not important in terms of the lifeline function, they do provide important sources of revenue for the harbor. Ships for specialized activities, such as fire fighting, can also be found at harbor facilities.

Materials that are transported through harbors can be classed as general cargo, bulk goods, specialty items, and passengers. General cargo is almost totally shipped by containers, that is, sealed boxes of various sizes. Bulk goods include petroleum products, coal and ore, cement, and granular food stuffs. These goods are typically transported in ships specifically designed for the particular bulk cargo. Specialty items such as automobiles, lumber and steel require their own unique considerations and methods of handling.

PORT FACILITIES

Before discussing the various types of construction employed in harbor facilities, the distinction between a pier and a wharf will be made. A pier is a structure, generally perpendicular to shore, for the berthing of vessels (Fig. 9.1). A wharf is a structure, generally parallel to shore, for berthing vessels (Fig. 9.1).

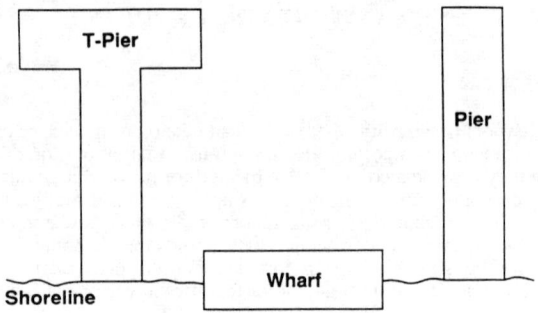

Fig. 9.1 Distinctions between wharves and piers.

Navigation channels are typically dredged from the floor of the harbor area to provide a safe transportation corridor for ships. These channels typically have sloped sides. Breakwaters and jetties are provided to limit the wave motion in a harbor area. Breakwaters and jetties typically employ rubble-mound construction (mound of stone or concrete units of different sizes and shapes either placed randomly or in courses) or cellular-steel sheet-pile construction. Concrete caissons are also used in breakwater construction. Shorelines in a harbor are protected from erosion by seawalls, quay walls, bulkheads, and revetments. These retaining structures are typically constructed with a concrete wall supported on piles protected by a sheet-pile cutoff wall and rubble to prevent loss of the foundation material. Rubble-mound construction, anchored vertical pile walls, gravity walls, cribs and cellular steel-pile structures are also used.

Piers and wharves may be classified as open or closed construction. In the open class, there are framed timber or reinforced concrete decks on piles or cylinders (Fig. 9.2). There are also relieving platform piers and wharves whose purpose is to provide the stability and cost savings that result from the use of shorter piles (Fig. 9.3). This type of pier can also accommodate buried utilities. For particularly wide piers (Fig. 9.4) and for wharves (Fig. 9.5), the open-type construction is used with a fill material, usually from dredging, that typically slopes from the dredged depth at the face of the pier to the deck elevation at the center of a pier or the rear of a wharf. This type of construction is often used as an economical means to support large loads, such as railroad cars, that are to be transported on the pier.

Closed type construction is more suited to the long faces of wharves, but this type of construction is sometimes utilized for piers. A cellular cofferdam, with interlocking steel sheet-piling, filled with granular material and having a concrete cap, is often used for this type of construction (Fig. 9.6). Closed construction wharves are also made with bulkheads of steel or concrete sheet-piling, with the earth pressures resisted by a tie-back and deadman system (Fig. 9.7), or resisted by battered piles (Fig. 9.8). Other structural systems used in closed construction include timber cribbing (Fig. 9.9), closed-cell wharves (Fig. 9.10), and massive gravity systems (Fig. 9.11). These three systems are in minimal use in the U.S., although older examples of timber cribs abound in many areas of the east and south.

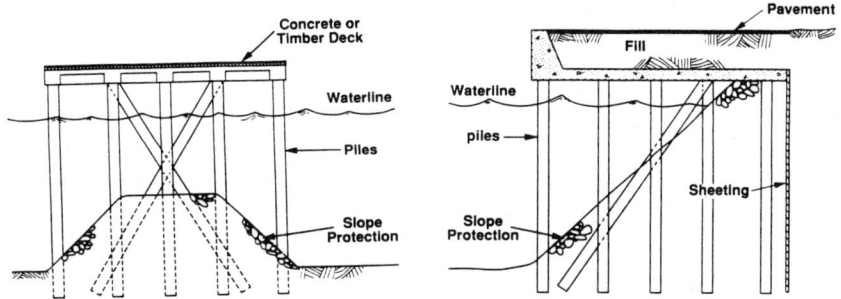

Fig. 9.2 Typical configuration of open pier. Fig. 9.3 Typical configuration of a relieving platform wharf.

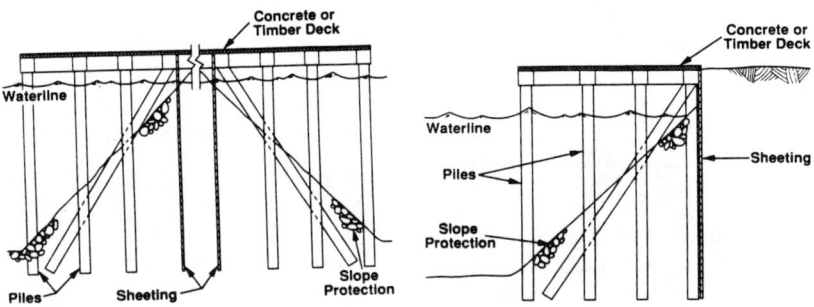

Fig. 9.4 Typical configuration of a relieving wide pier. Fig. 9.5 Typical configuration of a wharf.

In addition to the two primary types of construction used for piers and wharves, there are many piers that are engineered for special purposes. An example of such a facility are the piers that handle oil tankers. These piers consist of several mooring dolphins, a central hose-handling tower connected to the mooring facilities by catwalks, and a minimal facility to carry product lines to shore (Fig. 9.12). These facilities can be as simple as a single-point floating mooring to which the tanker is tied, with the liquid bulk cargo pumped to shore through a submarine pipeline.

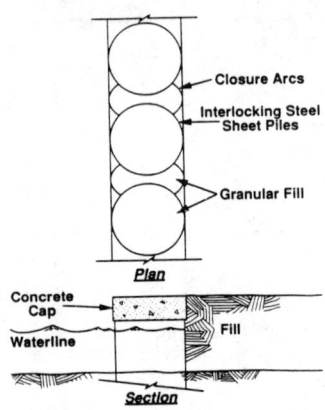
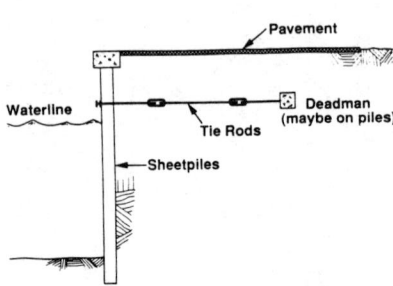

Fig. 9.6 Cellular cofferdam wharf construction.

Fig. 9.7 Ties sheet pile bulkhead.

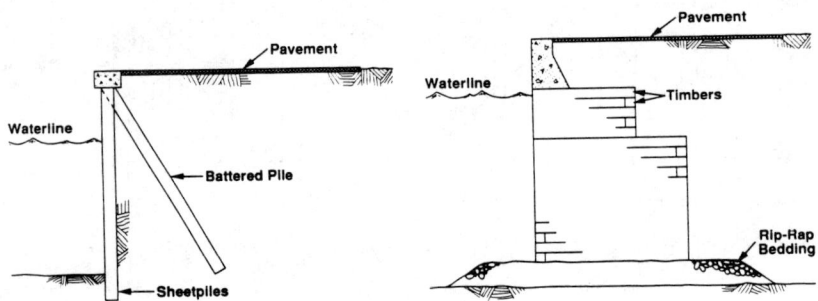

Fig. 9.8 Battered pile supported bulkhead.   Fig. 9.9 Timber crib wharf.

Piles are frequently used for pier and wharf support systems. In addition, fender piles are often located next to piers and wharves to prevent ships from colliding with the terminal facility. Piles are made of timber, steel, or concrete. They may be either vertical (plumb) or driven at some angle (battered) to resist lateral loading. Timber has

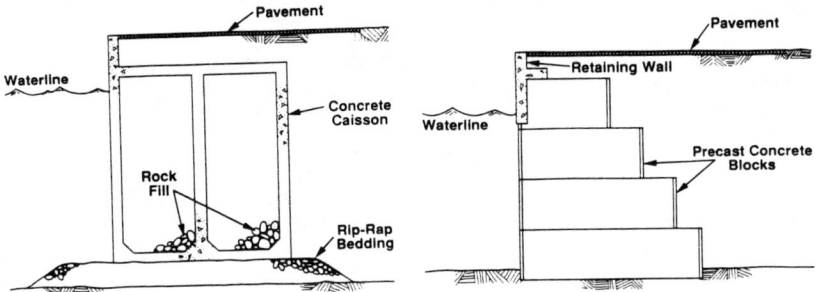

Fig. 9.10 Closed cell wharf.   Fig. 9.11 Precast gravity wall wharf.

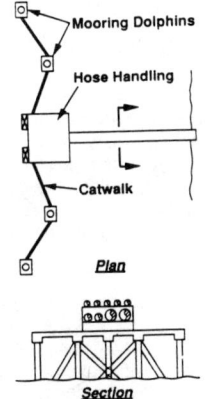

Fig. 9.12 Bulk liquid pier layout.

a tendency to be attacked by marine organisms and to break at the mudline or at the deck. Portions of timber piles that are in the splash zone have a tendency to rot. Steel H or wide flange sections that are typically used for piles are susceptible to corrosion, particularly in the splash zone. Precast and prestressed piles can be formed in various shapes and sizes but have a tendency to crack under certain load conditions. Piles may also be of composite construction such as concrete followers (extensions) on top of a timber base pile, concrete on top of steel, and steel or wood encased in concrete.

Specialized equipment for transporting cargo to and from a ship is located on the pier or wharf. This equipment includes wheel- and rail-mounted container cranes, ramps to unload automobiles and passengers, surface and overhead conveyor systems, stacker and reclaimer equipment for loading and unloading stockpiled cargo, rail spurs for trains, flexible rubber hoses or rigid pipes with rotating joints (chiksans) for transporting liquid bulk cargo, and underground utility lines.

Storage facilities include tank farms with berms to contain spills, open storage lots that are often behind dikes, sea walls or other retaining structures, container control towers, warehouses, and silos. Typically, 25-30 acres per berth are required for a cargo staging area. Inland transportation systems that serve the harbor consist primarily of roads and rail spurs, along with cargo transfer areas where goods are transferred from open storage lots to trains or trucks. Both storage facilities and transportation systems are similar to their inland counterparts. Information about earthquake response and post-earthquake investigation procedures for transportation systems can be found in other sections of these training notes.

TYPICAL EARTHQUAKE DAMAGE EXPERIENCED BY PORT FACILITIES

Table 1 (1) summarizes damage from various earthquakes that has been sustained by harbors. This table shows that all engineered components of a harbor are susceptible to damage from an earthquake. Damage has included excessive deformation of retaining structures, damage to piles, displacement of crane rails, settlement of pavements, submarine slope instability, damage to lifelines that interface with the ports, and damage resulting from tsunami.

From Table 1 it is evident that the most prominent source of earthquake damage to harbors is excessive porewater pressure buildup in saturated granular soils which lead to liquefaction, increased lateral pressures on retaining structures, and slope failure, including submarine landsliding. The liquefaction of soil can lead to large relative settlements of harbor facilities, particularly when one portion of a facility is supported on piles and the adjacent portion is supported on fill. Liquefaction and the associated soil settlement can also damage the inland transportation systems, such as railroad beds.

Damage directly related to vibration of the harbor structures has not been found to be as significant as damage that results from soil liquefaction. Seawalls and paved approaches to the piers have been damaged by ground vibration. Vibration of the pier can lead to excessive relative deformations, causing the deck to crack or buckle and causing piles to break at their connection to the deck. Battered piles have typically suffered more damage than vertical piles, since they provide most of the lateral stiffness, they obsorbe lateral loads. Cargo handling equipment, pipelines, and utility lines buried in the pier deck have also been damaged because of excessive relative deformations that result from vibration during the earthquake. Structures on the pier and adjacent storage facilities may be damaged by vibration in a similar manner as inland structures.

Broken water lines can result in increased settlement of the soil by washing out fine material. Also, the ability to fight fires that might occur after an earthquake can be hampered if water lines are broken. Breaks in natural gas lines have occurred at the shoreline/pier interface. Damage to these lines can prevent storage facilities from being heated and can cause all the normal problems associated with a gas leak. Loss of power will prevent normal harbor activities from taking place, particularly at night. Power is often turned off after an earthquake as a safety precaution against fires. Loss

Table 1
SUMMARY OF EARTHQUAKE - INDUCED DAMAGE TO PORT AND HARBOR FACILITIES

Earthquake			Port	Damage	
Location	Date	Magnitude	Location	Description	Possible Cause(s)
Kanto, Japan	09-01-23	8.2	Yokohama and Yokosuka	Concrete block quay walls: sliding, tilting, and/or collapse with some bearing capacity failure of rubble-stone foundation. Steel bridge pier: buckling of pile supports.	A C,E
Kitaizu, Japan	11-26-30	7.0	Shimizu	Caisson quay wall (183 m long): tilting, outward sliding (8.3 m) and settlement (1.6 m). L-Shaped block quay wall (750 m long): outward sliding (4.5 m) and settlement (1.2 m).	A,B,C
Shizouka, Japan	07-11-35	6.3	Shimizu	Caisson quay wall: outward sliding (5.5 m) and settlement (09. m) accompanied by anchor system failure	A,B,C
Tonankai, Japan	12-07-44	8.3	Yokkaichi Nagoya Osaka	Pile-supported concrete girder and deck: outward sliding (3.7 m) accompanied by extensive soil sliding. Sheet-pile bulkhead with platform: outward bulging (4 m). Steel sheet-pile bulkhead: outward building (3 m).	A,B,C
Nankai, Japan	12-21-46	8.1	Nagoya Yokkaichi Osaka Uno	Sheet-pile bulkhead with platform: outward bulging (4 m). Pile-supported concrete girder and deck: outward sliding 93.7 m). Steel sheet-pile bulkhead: outward bulging (3 m) and settlement. Gravity-type concrete block and caisson quay wall: seaward sliding (0.4 m) accompanied by soil sliding	A,B,C
Tokachi-Oki, Japan	03-04-52	8.1	Kushiro	Concrete caisson quay wall: tilting, outward sliding (6 m) and settlement (1 m).	A,B,C
Chili	05-22-60	8.4	Puerto Montt Talcuhauno	Concrete caisson quay walls: overturning and extensive tilting. Steel sheet-pile seawall: outward sliding (up to 1 m) and anchor failure. Gravity-type concrete seawall: complete overturning and sliding (1.5 m). Concrete block quay wall: outward tilting.	A,B,C A,B

147

Event	Date	Magnitude	Location	Description	Reference
Alaska	03-27-64	8.4	Anchorage	Dock structures: extensive seaward tilting with bowing, buckling, and yielding of pile supports.	B,D,E
			Valdez	Entire harbor: destroyed by massive submarine landslide.	
			Whittier	Pile-supported piers and docks: buckling, bending, and twisting of steel pile supports. Steel sheet-pile bulkhead: extensive bulging.	B,D
			Seward	Major portion of harbor destroyed by massive submarine landslide.	
Niigata, Japan	06-16-64	7.5	Niigata	Extensive damage due to liquefaction and sliding of soil strata. Summary of damage is as follows: Piers and landings: sliding (up to 5 m), submergence and tilting. Sheet-pile bulkheads: sliding (over 2 m), submergence, settlement (up to 1 m), and tilting. Extensive anchor failure. Quay-walls: seaward sliding (up to 3 m) and settlement (up to 4 m) with extensive anchor failure and wall tilting.	A,B,C
Tokachi-Oki, Japan	05-16-68	7.8	Hachinohe	Steel sheet-pile bulkheads: outward sliding (0.9 m), tilting, and settlement, with anchor failure.	A
			Aomori	Gravity-type quay wall: sliding and settlement (0.4 m).	A
			Hakodate	Gravity-type breakwater: sliding (0.9 m) and pavement settlement (0.9 m). Steel sheet-pile bulkhead: seaward tilting (0.6 m) and apron settlement (0.3 m). Quay-wall: settlement (0.6 m) and sliding (0.4 m).	A,B
Nemuro-Hanto-Oki, Japan	06-17-73	7.4	Hanasaki	Gravity-type quay wall: sliding (1.2 m) and settlement (0.3 m) with corresponding apron settlement (1.2 m). Steel sheet-pile bulkhead: sliding (2 m) and anchor failure.	A,B
			Kiritappu	Gravity-type quay wall: relatively minor damage.	
Miyagi-Ken-Oki, Japan	06-12-78	7.4	Shiogama	Concrete gravity-type quay wall: outward tilting (0.6 m) and apron pavement settlement (0.4 m).	A,B
			Ishinomaki	Steel sheet-pile bulkheads: outward sliding (up to 1.2 m) and apron settlement (up to 1 m). Concrete block retaining wall: sliding, tilting, and cracking with corresponding pavement settlement (0.2 m) relative to wall.	
			Yuriage	Concrete block gravity quay wall and steel sheet-pile bulkhead: large horizontal displacements (up to 1.2 m).	
			Sendai	Steel sheet-pile bulkheads: cracking and settlement of apron and pavements.	

Location	Date	Magnitude		Description	Causes
Offshore, Central Chile	03-03-85	7.8	San Antonio	Concrete block seawall at western extremity of port: collapse of about 60 percent of seawall length, leading to significant tilting of crane structures.	A,B
			Valparaiso	Seawalls at Sites 6 and 7 (at west side of port): several inches of soil movement at top of wall, causing movement of crane rails and disruption of crane operations. Seawall at passenger terminal: lateral movement of seawall, settlement of backfill.	
Loma Prieta, CA	10-17-89	7.1	Redwood City	Broken waterlines, damaged batter piles, separation of concrete ramp from pier at Wharf 1. Temporary disruption of power and communications. No disruption of port operations.	A,B
			Richmond	Ruptured gasoline storage tank at UNOCAL terminal. Broken water lines at Ford plant east of Terminal 3. Port fully functional two days after earthquake.	A,B
			San Francisco	Settlement of portions of piers supported by fills (rather than by piles). Significant structural damage to various buildings along waterfront, including clock tower at Ferry Building. Many broken water lines, broken batter piles, cracked concrete decks above piers. Damage to container crane.	A,B,E
			Oakland	Seventh Street Terminal: 1 ft. of subsidence and 4" - 6" lateral movement of sand fill dike. Settlement of crane rail, broken batter piles, punching of piles through deck. Extended closure of terminal. Matson Terminal: separation between pile and bottom of slab, cracking of piles, wharf stayed in operation but with limited live loads while crane was operating. Middle Harbor Terminal: 6" - 24" of subsidence in yard area. Howard Terminal: 4" outward movement of landside crane rail (although crane was operational after earthquake). Broken concrete piles and 4" - 6" of subsidence beneath old shed area. General: many water lines broken within port.	A,B,C
Cabanatuan Philippines	7-16-90	7.7	San Fernando	Damage to battered and vertical piles, separation of concrete approach ramp from pier, extensive cracking of concrete pavement leading to the pier, excessive displacement of the concrete deck causing cracking of the concrete and fracture of the rebar, damage to conveyors and pipelines from excessive displacements of the pier. Cracks in the seawall.	E

Legend

A: Excessive lateral pressure from backfill materials, in the absence of complete liquefaction, and possibly accompanied by reduction in water pressure on outside of wall.
B: Liquefaction
C: Localized sliding
D: Massive submarine sliding
E: Vibrations of structure

of communication systems has also been noted at harbor facilities during past earthquakes. However, loss of communications has not had the adverse effects on the harbor activities as has the damage to the other lifelines mentioned above.

WHAT TO LOOK FOR

Examples of damage that have been documented at harbors from previous earthquakes are shown in this section. These examples are provided to give the investigator an idea of what to look for when conducting a post-earthquake investigation.

Figure 9.13 shows damage to a seawall at the Port of San Fernando in the Philippines. The investigator should note the general pattern of the cracks and the quality of construction of undamaged portions seawall.

Figures 9.14 and 9.15 show damage to reinforced concrete piles at the Port of Oakland. Most damage that is visible will be at the pile/deck interface. It may be possible to examine the bases of piles close to shore during low tide. The investigator should examine the exposed rebar for excessive corrosion to determine if the piles were in a damaged state prior to the earthquake. Liquefaction of fill under a pier can be identified by sand boils as seen in Figure 9.16. Differential settlement of the portions of the pier that are supported on fill relative to the portions of the pier supported on piles is shown in Figures 9.17 and 9.18. The investigator should try to obtain information from the port engineers concerning the construction procedures used in placing the fill. As an example, was the fill vibro-compacted to minimize the settlement?

Fig. 9.13 Cracks in the seawall at the port of San Fernando

Fig. 9.14 Damage to a battered pile at the Port of Oakland.

Fig. 9.15 Damage to a vertical pile at the Port of Oakland.

Fig. 9.16. Sand boils observed at the Port of San Francisco.

Fig. 9.17. Settlement of an asphalt deck caused by liquefaction of the underlying fill observed at the Port of San Francisco.

Fig. 9.18. Settlement of an asphalt deck caused by liquefaction of the underlying fill observed at the Port of Oakland.

Figures 9.19 through 9.22 show damage to concrete decks that resulted from lateral spreading at the Port of San Fernando. The temporary repairs used to fix these types of damage should be noted by the investigator.

Container cranes can become inoperable because of damage to the crane itself or because of damage to its rails. Figure 9.23 shows differential settlement of the inboard rail (supported on fill) relative to the outboard rail (supported by a beam on piles) at the Port of Oakland. Figure 9.24 shows damage to a conveyor at the Port of San Fernando that resulted from longitudinal deformation of the pier due to lateral spreading. Figure 9.25 shows the excessive relative displacement of a pipeline at the Port of San Fernando. This damage was caused by the concrete approach span collapsing on the pipeline.

Cargo storage areas can be damaged directly by ground vibration or by liquefaction of the underlying material. Figure 9.26 shows damage to the walls of a warehouse at the Port of San Francisco that has resulted from settlement of the underlying fill material. Figure 9.27 shows the settlement of a asphalt lot used for container storage at the Port of Oakland. When settlement like this is observed, the investigator should try to determine if it is the result of liquefaction or if it was caused by breaks in underground water lines.

Figures 28 through 31 show damage to inland transportation systems. Lateral and vertical displacements of a railroad at the Port of Oakland are shown in Figs. 9.28 and 9.29. This damage was caused by displacements of the railroad supported on fill relative to the portions of the railroad supported by the bridge that is in turn supported

153

Fig. 9.19 Buckling of a concrete deck at the Port of San Fernando.

Fig. 9.20 Collapsed concrete approach span at the Port of San Fernando.

Fig. 9.21 Damage to concrete deck due to lateral spreading at the Port of San Fernando.

Fig. 9.22 Relative motion between the main pier and a small side pier opens gap in deck at the Port of San Fernando.

Fig. 9.23 Differential settlement of crane rails at the Port of Oakland.

Fig. 9.24 Damaged conveyor at the Port of San Fernando.

Fig. 9.25 Damaged pipeline at the Port of San Fernando.

Fig. 9.26 Damage to a warehouse wall caused by settlement of the underlying fill at the Port of San Francisco.

Fig. 9.27 Settlement of an asphalt container storage lot at the Port of Oakland.

Fig. 9.28 Vertical displacement of a railroad at the Port of Oakland.

Fig. 9.29 Lateral displacement of a railroad at the Port of Oakland.

by piles. Minor settlement and lateral spreading of a roadway next to a seawall at the Port of Oakland is shown in Fig. 9.30. Damage to the concrete road leading to a pier at the Port of San Fernando is shown in Fig. 9.31.

CASE STUDY - Damage Sustained By Ports In The San Francisco Bay Area As A Result Of The 1989 Loma Prieta Earthquake (2)

INTRODUCTION

This report summarizes the post-earthquake investigation of the major seaports in the San Francisco Bay area. The investigations of the ports of San Francisco, Richmond, and Redwood City were made on November 20 and 21, 1989. An

Fig. 9.30 Damage to a roadway at the Port of Oakland

Fig. 9.31 Damage to the concrete roadway leading to a pier at the Port of San Fernando

investigation of the Port of Oakland was made on November 28, 1989. Figure 9.32 shows the location of these ports.

During the investigation an attempt was made to both interview the engineers responsible for the maintenance and operation of the facilities, and inspect the actual damage to the facilities so that the following information could be obtained:

Description of each port facility
Description of earthquake damage
Description of the damage's impact on users of each port
Description of methods to restore service
Description of obstacles to the restoration of services
Effects of damage to other lifelines on the operations of each port
Description of emergency plans that each port facility had and how useful they were

PORT OF REDWOOD CITY

Description of Facilities

The Port of Redwood City is located at the southern end of San Francisco Bay, on the western side where Redwood Creek enters into the Bay as shown in Figure 9.32. This was the port nearest the epicenter of the earthquake and the smallest port facility that was visited. Figure 9.33 shows a layout of this facility. Cement, scrap metal, and salt are the primary goods that are shipped from this port. Four-hundred eighty-eight thousand metric tons of freight were handled by the Port in 1988. Tanks for petroleum products are also located at this port, along with a public marina and restaurants. The Port was constructed on bay mud, with the wharfs supported by both concrete and wood piles. The Port of Redwood City is responsible for all the maintenance of its facilities and has a one-man, in-house maintenance crew. All the commercial wharfs are accessible by both rail spurs and roadways.

Earthquake Damage

Although this Port was located closest to the epicenter of the earthquake, it sustained the least damage of the Bay area ports. Total damage was estimated at $260,000 and consisted primarily of broken water lines, damaged battered piles and separation of the concrete ramp at Wharf 1. This port was never closed and the earthquake damage had no adverse effect on the operation of the facility. Water lines were repaired within one week after the earthquake. Other repairs to the Port will be contracted out.

There was only a temporary interruption in electrical and telephone service. Both services were restored within eight hours after the earthquake. Traffic was jammed on the major highways that service the Port for most of the evening of the earthquake. No effects from damage to other lifelines were reported.

Earthquake Response Plans

This port had no formal emergency response plan. The business manager stated that they do not intend to develop such a plan at this time.

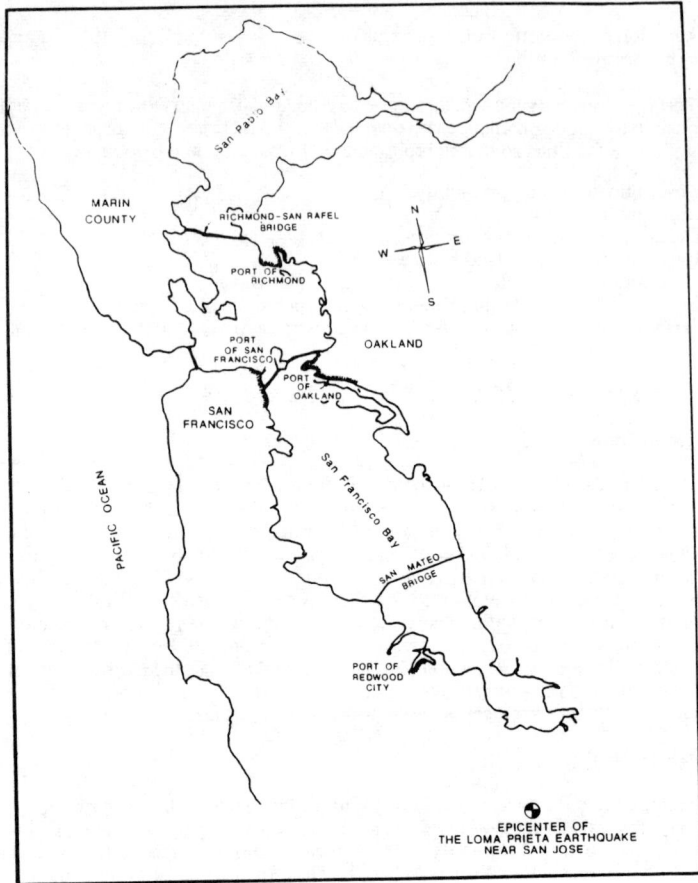

Fig. 9.32 Layout of ports of San Francisco Bay

PORT OF RICHMOND

Description of Facilities

The Port of Richmond is located at the northeast end of San Francisco Bay near the entrance to San Pablo Bay. This port handles the most cargo (18 million short tons per year) of any in the Bay area and is the primary Bay area port for petroleum products and liquid bulk cargos. The majority of the terminals are privately owned. Other port properties are leased to tenants with various maintenance agreements. Figure 9.34 shows a map of the port facility along with a summary of the various activities, owners, and tenants at those facilities. Portions of this port are constructed on a

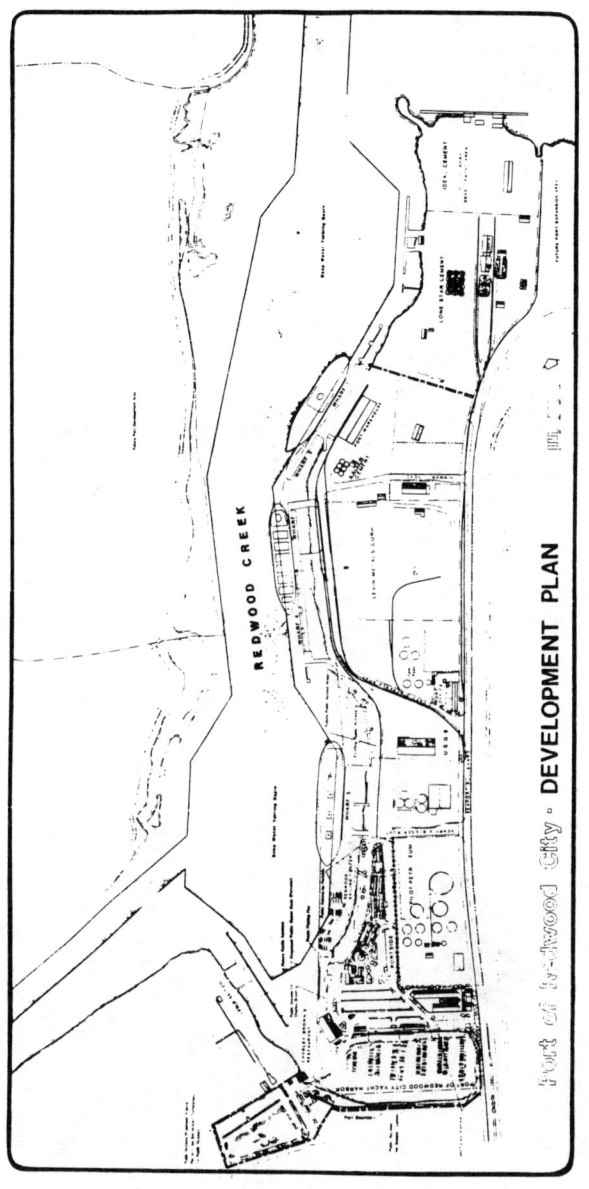

Fig. 9.33 Layout of Port of Redwood City

Fig. 9.34 Layout of Port of Richmond

rock-site and other portions are on fill. The Port is maintained by a staff of two engineers and one maintenance man.

Earthquake Damage

The primary damage sustained by this port was a ruptured gasoline storage tank at the UNOCAL terminal (Terminal No. 13 in Fig. 9.34). Fuel from this tank was contained by the surrounding berm. Some liquefaction was reported at the undeveloped areas on fill south of terminal No. 3. Sand boils were noticed at Terminal No. 3 and broken water lines were reported at the Ford plant that is located on fill east of Terminal No. 3.

Unloading of cargo was delayed for 24 hours because of the gasoline leak. The fire department shut off power to the terminals because of a fear that sparks might ignite the fumes from the spilled gas. Also, a ship that had arrived to unload automobiles was sent to deep anchorage until the gasoline was cleaned up. The other damage that was reported had no effect on the operation of facilities at the terminals where it occurred.

Because damage was sustained exclusively by privately-owned terminals the Port of Richmond was not involved in restoration of services. Within 48 hours after the earthquake the entire port facility was functioning as normal. There was no damage to other lifeline services other than the intentional interruption in electrical service previously discussed. To help with the commuter problems that resulted from the closing of the Bay Bridge, the Port of Richmond added facilities for additional ferries at terminal No. 3. These facilities were operational on Monday, October 23 and would have been operational on Thursday, October 19, but the low free-board of the ferries relative to freight vessels that are normally moored at this terminal required the addition of a floating dock facility. Prior to the re-opening of the Bay Bridge, the new ferry service transported 400-500 passengers a day the 8 miles to the Ferry Building at the Port of San Francisco in approximately 45 minutes.

Emergency Response Plans

Engineers at the Port of Richmond said that they have no emergency plans for earthquake disasters because the Port is under jurisdiction of the City of Richmond and the City has charged the Fire Department with emergency response.

PORT OF SAN FRANCISCO

Description of Facilities

The Port of San Francisco extends for seven miles along the west side of San Francisco Bay from west of Aquatic Park to India Basin as shown in Fig. 9.32. Activities at this port are diverse and include import and export of general cargo primarily in container form, ship repair, passenger and cruise ships, a fishing industry and tourist attractions. Three million metric revenue tons of general cargo pass through this port in a year. Annual revenues from the Port are estimated at 38 million dollars. Figure 9.35 shows that most of the San Francisco waterfront is constructed on fill. Most piers in the Port were constructed on fill behind seawalls with the portions of the piers extending past the seawalls supported on piles. Concrete, timber and timber encased in concrete piles are used throughout the Port. The central portions of the piers

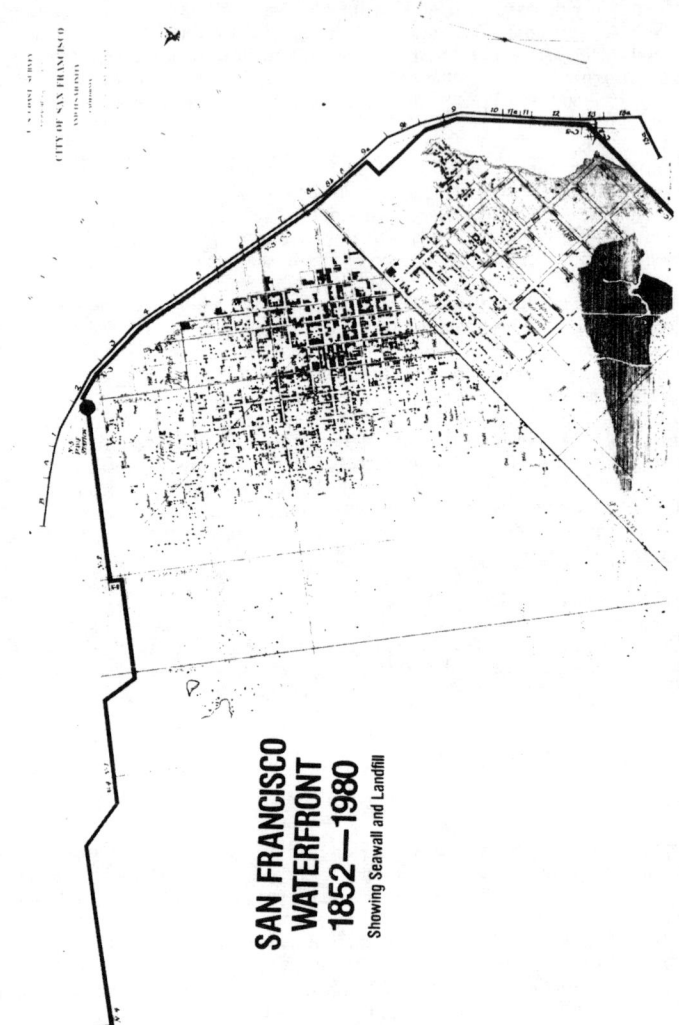

Fig. 9.35 Port of San Francisco and land fill areas

were built on fill to accommodate rail spurs, however, today the piers are serviced primarily by trucks. The Port of San Francisco has a 11-man engineering staff, four inspectors and a maintenance crew of approximately 70 people.

Earthquake Damage

Engineers at the Port estimate that repairs to the Loma Prieta Earthquake damage will cost 8-10 million dollars. Tenants suffered more damage for which the port is not responsible. As with the other ports in the Bay area, the tenants responsibility for repair and maintenance of their Port facility differs with different lease agreements. The primary cause of damage was liquefaction of the fill material that resulted in settlement of the portions of the piers supported by fill relative to the portions of the piers supported by piles. Settlement was observed to continue for several weeks after the earthquake. Sand boils that result from the upward flow of water caused by the excess pore water pressure associated with liquefaction could be observed at Pier 45. Significant structural damage that resulted in buildings being condemned occurred at different locations along the waterfront. Structural damage included cracked concrete walls and displaced asphalt decks in warehouses on Pier 45 and 48 that resulted from settlement of the underlying fill, cracking and collapse of unreinforced clay tile walls in an office building on Pier 70, buckling of columns in the clock tower above the Ferry Building (the offices for the Port of San Francisco are located in the Ferry Building). The Agriculture Building located adjacent to the Ferry Building was closed because of differential settlement. Other damage at the Port included many broken water lines, many broken batter piles, cracked concrete decks above piers, and damage to five container cranes.

Although it had limited effect on the operation of the Port, significant damage was sustained by the seventeen-story clock tower above the Ferry Building. Both the clock tower and the Ferry Building had survived the 1906 earthquake. The clock tower's primary structural elements are four laced columns. Each column consists of two channels sections tied together with riveted plate laces on either side. Diagonal tie-rods with turnbuckles are used to provide lateral stability. At the thirteenth level two of the columns had buckled. The damaged sections have been cut out and temporary splices have been added. At several levels the tie-rods had noticeable deformation but without knowing the condition prior to the earthquake it is difficult to tell if this was a result of the Loma Prieta Earthquake.

Commercial freight operations were mostly unaffected by the earthquake damage. Two damaged container cranes were jacked back onto their rails and the damaged steel plates on the crane's legs were replaced by a contractor. While this was being done stand-by cranes were brought into service and ships were being unloaded by Thursday, October 19. Commercial fishing was probably the most adversely affected business because of the need to close two warehouses on Pier 45 that were used for transportation and processing of fish. Grain elevators near the south container terminal had to be closed until damage to a drag conveyor bridge structure could be repaired. The Port of San Francisco lost revenues from the businesses in the tourist sections of the Port that, as part of their lease agreements, pay the Port a percentage of their profits. These businesses continue to suffer from the reduced tourism that has resulted from the earthquake.

Water and gas lines were repaired within one week after the earthquake. The Port of San Francisco did not have enough plumbers to accomplish the repairs that fast, but additional plumbers were provided by a private contractor and by the City of San Francisco. Temporary fixes to the differential settlement problems were handled by constructing asphalt ramps over cracks in the existing pavement. These repairs were done by in-house maintenance people and there was no difficulty getting repair material. In buildings that had significant structural damage offices were moved to other locations. Fish processing, that was normally handled in a warehouse on Pier 45, was moved to Pier 33.

Telephone service was never out at the Ferry Building. however, the lines were jammed because of the volume of use. Loss of electricity to Pier 45 caused problems for the fish processing business as these businesses were required to rent generators to prevent thawing of their freezers. Power had been turned off by PG&E because of concern for fires associated with gas leaks. Electric power was restored by 2:00 p.m. on October 19. Damage to the Bay Bridge resulted in an increased use of ferries by commuters and additional temporary ferry terminals were set up at Pier 1. These terminals were operational on Monday, October 23.

Emergency Response Plans

The engineers at the Port of San Francisco said that they have no formal emergency plans for contending with earthquakes and at this point they do not plan to develop any. The city-wide emergency plan includes the Port. Utility shut-off manuals have been distributed to maintenance personnel and local fire stations.

A fire boat that is under the jurisdiction of the Port and that the City of San Francisco was previously planning to put out of service as a cost saving measure was instrumental in putting out the fires in the Marina district.

PORT OF OAKLAND

Description of Facilities

The Port of Oakland occupies 19 miles of shore on the east side of San Francisco Bay as shown in Fig. 9.32. A detailed layout of the Port is shown in Fig. 9.36. Oakland International Airport is also under the jurisdiction of the Port of Oakland, however, this report will focus exclusively on the sea port. Container freight is the primary item handled by this port. There are currently 25 container cranes that handle 14 million metric revenue tons of cargo annually. Most of the Port is constructed on fill behind seawalls with wharfs supported on piles extending beyond the seawalls. All the terminals are served by rail spurs and trucks. Some of the terminals are leased to private concerns with varying maintenance agreements while other terminals are used as needed by customers with the Port providing the maintenance. The Port is maintained by an engineering staff of five people and a 70-person maintenance crew.

Earthquake Damage

It was initially estimated that 75 million dollars will be required to repair the damage caused by the Loma Prieta Earthquake. Damage occurred primarily at the Seventh St. Complex and the Middle Harbor. The Inner and Outer Harbors were relatively undamaged. As with the Port of San Francisco, the primary cause of damage was

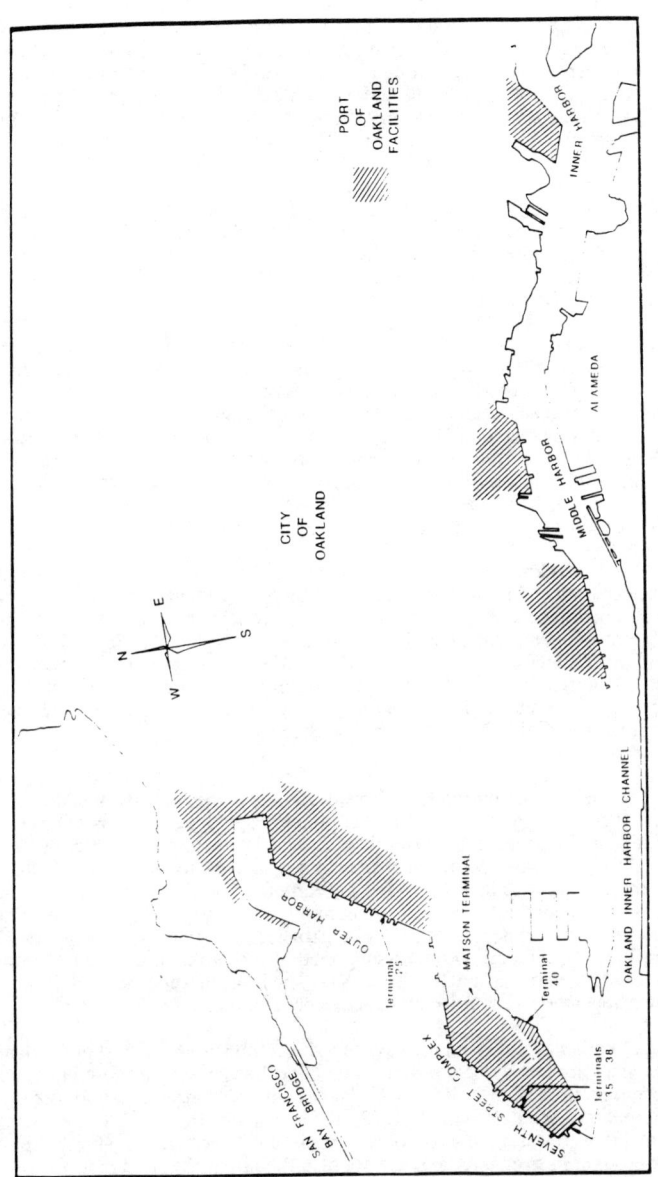

Fig. 9.36 Port of Oakland and facilities

liquefaction of the fill and the resulting settlement and spreading of areas on fill relative to areas supported by piles. There were numerous broken water and fire lines that washed the fine material from the soil causing both settlement and uplift of the asphalt pavement at numerous locations throughout the Port. Concrete batter piles were broken at the Seventh St. Complex terminals and at the Matson Terminal. Figure 9.37 shows a cross-section of these terminals. The piles support a concrete deck covered by three feet of sand through which utility lines run. Three inches of asphalt pavement cover the sand. The damaged piles at this terminal resulted from the inertial forces generated by the large mass supported on top of these piles. At Terminal 25 of the outer Harbor horizontal ground acceleration was measured at 0.29 g's and vertical ground acceleration was measured at 0.07 g's. Horizontal acceleration on the wharf at Terminal 25 was measured at 0.45 g's. Wood piles with concrete followers were broken at the Middle Harbor resulting in a building that was supported on these piles being condemned. These piles broke at the wood-concrete interface. Throughout the port damage to the piles and settlement of the fill occurred up to three weeks after the earthquake. A container crane derailed at the Middle Harbor and the 100-ft gauge distance between the rails increased up to four inches. Throughout the Port damage was sustained by 23 container cranes and was estimated at 500,000 dollars. At the Seventh Street terminals the crane rails that were on fill settled as much as 12-15 inches relative to the rail on piles rendering the container cranes inoperable . The rail spur serving Terminal 40 is currently out of service because of horizontal and vertical displacement of the rails on fill relative to a portion of the spur supported by piles. Significant settlement and separation of the truck access road to Terminals 35-38 at the Seventh St. Complex occurred as a result of liquefaction.

The Seventh St. Complex is currently shut down. Half of this complex will be temporarily repaired and back in service by February. Permanent repairs to the other half are scheduled to be completed by June. At that time work will begin to make the other temporary repairs permanent. Except for the crane at the Middle Harbor that derailed, all other container cranes remained operable. The derailed crane was repaired and operable within a day after the earthquake despite the increased gauge length of the rails. Although settlement and uplift of the pavement occurred throughout the staging areas for container freight, this damage did not have an adverse effect on the overall operation of these facilities.

Repairs to the Port are currently underway and are being done by outside contractors. Temporary repairs consisted of placing asphalt at locations of large discontinuities in the pavement. Water lines that the Port was responsible for were repaired within two weeks. Water lines maintained by the City of Oakland that served the Port were repaired within a week after the earthquake. Permanent repairs to the crane rails at the Seventh St. Complex are currently underway. These repairs entail the removal of the settled crane rail originally constructed on a grade beam supported by fill. The rail is to be replaced with one supported by a beam on piles and tied into the wharf to provide further earthquake resistance. The Matson Terminal that does its own maintenance appeared to be replacing the pavement on most of its property.

Damage to other lifelines did not have any direct impact on the Port of Oakland. Power was lost temporarily. No problems with gas lines occurred. Fire lines ruptured but there were no fires hence this damage did not adversely affect the Port. One obstacle to repair of the Port is the lack of reliable benchmarks for surveying. Because the benchmarks were assumed to move as a result of the earthquake, a level circuit will have to be brought in from a remote site not disturbed by the ground motion.

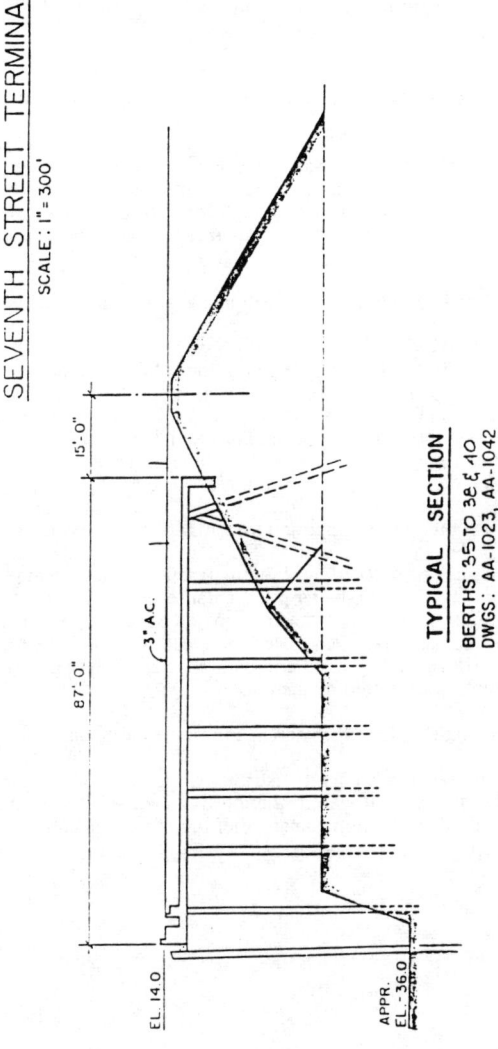

Fig. 9.37 Typical section of Oakland Terminal

Emergency Response Plans

The Port had no formal emergency response plan but they did have an emergency call list of contractors and equipment suppliers. A formal emergency response manual is planned for the future but is not currently in the development phase.

CONCLUSIONS

- Extensive damage to port facilities can be expected from settlement and soil liquefaction.

  The construction of ports in the San Francisco Bay area predates modern engineered fill so that the Loma Prieta Earthquake caused extensive damage from differential settlement and soil liquefaction. Much more extensive and severe problems can be expected in future earthquakes if ground motions are more severe or last longer.

- The use of different foundation types within a given structure increases the potential for damage.

  Differential settlement of pile-supported portions of the terminal relative to the portions supported by fill caused structural damage to buildings and equipment that straddle the different foundation systems. Container cranes and their rails are particularly susceptible to damage from differential settlement. Future port designs should have a consistent foundation system under a structure and structures should not be supported directly on fill or by foundations on fill.

- Unnecessary weight on pile supported systems should be avoided.

  Damage to piles can be reduced by avoiding terminal designs that place excessive mass on top of the piles. The batter piles under terminal 40 at the Port of Oakland that support a three-foot layer of sand on top of a concrete slab were severely damaged. This damage could have been mitigated if an alternate method of running utility lines was developed that did not require the sand layer thus reducing the mass (and inertial forces) supported by the piles.

- Improved planning should reduce damage in future earthquakes.

  The Port of San Francisco's plan to distribute utility shut-off manuals to maintenance personnel and local fire stations is a inexpensive precaution that could prevent additional damage immediately after future earthquakes.

9. TRANSPORTATION - HIGHWAYS

Elements of a highway transportation lifelines can include bridges, roadways, and tunnels. This section summarizes principal sources of damage to these elements during past earthquakes, and what aspects of these elements that should be scrutinized during post-earthquake inspection.

Damage to highways systems have been very disruptive to post-earthquake response operations. The most common causes of highway system damage from earthquakes has been landslides that cover or sweep away roadbeds, liquefactions and lateral spreading that damages roadbeds and causes severe bridge damage or collapse, and vibration effects on bridge deck supports. Damage to bridges can be particularly disruptive since repair time can be lengthy and alterative traffic flow is generally difficult to accommodate, particularly at water crossings.

The following discussion relates to the detailed performance of bridges, roads, and tunnels.

HIGHWAY BRIDGES

Typically bridges include single or multiple spans as well as continuous/monolithic spans. Bridges may be straight or skewed, fixed, movable (draw bridge, or rotating, etc.) or floating. Reinforced concrete is the most common construction material while steel, masonry, and wood construction are common at water crossings. Typical foundation systems include abutments, spread footings, battered and vertical pile groups, single-column drilled piers, and pile bent foundations. Bents may consist of single or multiple columns, or a pier wall. The superstructure typically comprises girders and deck slabs. Fixed (translation prevented, rotation permitted) and expansion (translation and rotation permitted) bearings of various types are used for girder support to accommodate temperature and shrinkage movements. Shear keys are typically used to resist transverse loads at abutments. Abutment fills are mobilized during an earthquake as a bridge moves into the fill (longitudinal direction), causing passive soil pressure to occur on the abutment wall.

Bridge damage can occur in the superstructure, the substructure, including the foundation, or in the abutment and approaches.

Superstructure Damage

The collapse of the superstructure will result in lengthy and costly repairs. Superstructure damage generally occurs because of a collapse of the girders, failure of anchor bolts, cracking and spalling of concrete girders and deck slabs, and other forms of local buckling or distress. Collapse of the girders is the most severe form of superstructure damage, and this may be caused by a lack of continuity in the superstructure, inadequate support length for girders, or gross movement at the pier foundation or abutment which are generally attributable to some form of soil failure.

The most vulnerable components of a bridge include support bearings, abutments, piers, footings and foundations. A common deficiency is that unrestrained expansion joints are not equipped to handle large relative displacements (inadequate support length), and simple bridge spans fall. Skewed bridges in particular have performed poorly in past earthquake because they respond partly in rotation, resulting in an unequal distribution of forces to bearings and supports. Rocker bearings have proven most vulnerable. Roller bearings generally remain stable in earthquakes, except they may become misaligned and horizontally displaced. Elastomeric bearing pads are relatively stable although they have been known to "walk out" under severe shaking.

Substructure Damage

Substructure damage generally manifests itself in the form of damage to columns, abutments, or foundation piles or footings. Column damage can be caused by shear of flexural failures, anchorage failure of longitudinal reinforcement, or insufficient confinement at plastic hinge locations. These types of failure may cause the collapse of the superstructure.

Foundation Failures

Foundation failures due to excessive ground deformations, and/or loss of stability and bearing capacity of the foundation soils. This has caused tilting, settlement, sliding, and overturning of the bridge substructures, which has often led to severe cracking or complete failure of these substructure elements. These large displacements may also cause superstructure damage and failure of bearing supports.

Liquefaction has been associated with many bridge failures, particularly at river crossings. Liquefaction can cause the abutments to move towards each other pushing the span off of intermediate supports. Intermediate supports can move, tip or settle. The motion of girder supports coupled with a lack of continuity of the superstructure or inadequate support length for the girders can lead to the collapse of the superstructure.

Abutment and Apporach Damage

Liquefaction and lateral spreading, particularly at river crossings, tends to cause the abutment to move towards the center of the river. Loads applied to the deck support system may cause piers to fail, or cause Excessive abutment forces from the backfill can be in phase with the seismic inertia force from the superstructure, causing superstructure damage or failure, separation of the abutment wing walls from the end walls, and settlement of back fill materials.

Settlement of the roadbed just behind the abutment is very common. This can be caused by subsidence of fill behind the abutment, spreading of the soil behind the abutment, liquefaction, or the failure of the abutment. This settlement can grow over a period of a week after the earthquake and can exceed 30 cm. While temporary repairs are easy to make, large numbers of these failures can be very disruptive.

Post-earthquake Inspection of Bridges

Post-earthquake inspection of highway bridges should focus on: (1) observation of signs of settlement, tilting, heaving, or sliding of bridge foundations; (2) any signs of cracking, spalling or pounding of abutment wall elements or settlement of backfill materials; (3) any signs of liquefaction (ie, sandboils) of adjacent soil and backfill materials which could lead to foundation damage or excessive abutment forces; (4)

signs of rubbing or sliding at bridge expansion joints (even those joints with cable restrainers); (5) signs of shear cracking or flexural cracking in reinforced concrete columns, beams, and beam-column joints; (6) signs of excessive deformation and cracking/fracture of bolted connections. If possible, determine the type of foundations that are used. While this may be possible by inspection, more detailed information can be obtained from highway departments. Also available from highway departments would be bridge and highway design specifications related to seismic loads.

ROADWAYS

Roadways include pavements, embankments, and cut sections. Pavements include a base and subbase. Pavement types may be either portland cement concrete or asphaltic concrete. Base and subbase materials include aggregate, cement treated aggregate, and lime-stabilized, bituminous, and soil cement bases. Embankments and cut sections may or may not include retaining walls.

Roadway damage can result from failure of the roadbed or failure of an embankment adjacent to the road. Roadbed damage can take the form of soil slumping under the pavement, and settling, cracking, or heaving or pavement. Embankment failure may occur in combination with liquefaction, slope failure, or failure of retaining walls. Such damage is manifested by misalignment, cracking of the road surface, local uplift, subsidence, buckling, and blockage of the roadway. sloping fills where compaction is commonly poor are particularly vulnerable to slope failure.

Since roadways have an extended length, it increases the probability that they could encounter regions that are susceptible to earthquake-induced geologic hazards. Such hazards -- which can include surface fault rupture, landslides, and liquefaction, lateral spreading, settlement, and heaving of the subsurface soil materials -- can cause severe damage and loss of function to the roadway element. Several of these hazards are particularly severe for roads that run adjacent to rivers and small streams. Therefore, observation of any signs of these potential hazards should be emphasized during post-earthquake inspection of roadways.

TUNNELS

In general, tunnels pass through alluvium or rock, or may be of cut and cover construction. Tunnels may be lined, or unlined, and may be at any depth below the ground surface. Lining materials include brick and both reinforced and unreinforced concrete. Heavy timbers and wood lagging (grouted and ungrouted) may also be used to support tunnel walls and ceilings. Tunnels may change in shape and/or construction material over their lengths. Tunnels lengths may range from less than 30 m to several km.

Experience from past earthquakes has shown that tunnels are relatively insensitive to effects of vibratory ground shaking, because of their confinement by the surrounding soil or rock material. They are more prone to damage from: (l) large ground deformation or failure due to liquefaction, sliding, and subsidence of soil materials or spalling, cracking, and block motion in rock materials (particularly at portals and in shallow excavations); (2) rupture of any fault that may intersect the tunnel; and (3) landslides or rock fall at tunnel entrances. Specifically, damage has been noted at tunnel weak spots, such as intersections, bends, changes in shape, construction materials, or soil conditions.

Accordingly, post-earthquake inspection of tunnels should focus on observing any signs of (1) liquefaction or excessive movement of the surrounding soil materials; (2) soil or rock movement at portal openings; and (3) excessive cracking, spalling, and/or deformation of the tunnel walls, particularly at locations of stress concentrations (eg at tunnel intersections, end walls at station-tunnel junctions, or crossings through junctions of substantially different geologic media) or at active fault crossings.

REFERENCES

1. "Seismic Risk to Port and Harbor Facilities," in the Proceedings of the Workshop on Seismic Issues Associated With Construction of the 2020 Project, Port of Los Angeles, San Pedro, California (May, 1990).

2. "Lifelines - Ports," Earthquake Spectra, Supplement to Vol. 6, May 1990, pp. 282-293.

AIRPORT CHECK LIST

___ Basic Data: No. and length of runways, passenger and freight traffic, map

Airfield-Side

___ Control Tower: power, windows, equipment, ceiling, radio communication, telephone communications
___ Air Traffic Control: radars, beacons, approach lights, ground control radios
___ Runway and taxiway damage
___ Fuel Supply: Tanks, power, piping
___ Status of aircraft service facilities, freight facilities, aircraft food service, freight storage facilities
___ Communications: radio saturation, repeater operability, emergency power, telephone operation, phone saturation, on-feild and off-field communications
___ Passenger Baggage Facilities
___ Emergency response facilities: power, security, fire, medical

Land-Side

___ Terminal: structure, power, jetways, non-structural damage: ceilings, fire suppression,
___ Communications: public address system, airport service phones(courtesy, security, maintenance), on- and off- field telephone communications, contact with emergency response, operation of 911, power
___ Airport based transportation systems
___ Airport-community transportation links
___ Emergency power

General
___ Water, sewage, and commercial power systems

Impacts of Operations

___ Flag-offs and diversions
___ Personnel problems
___ Passengers
___ Evacuations

PORT CHECK LIST

General Information
    Names, addresses, phone numbers, and titles of contact persons.
    Organizational management (private, municipal, district, etc.).
    Type and amount of cargo handled by the harbor, annual revenues.
    If ferries use the facility, the number of passengers per day before and after the earthquake.
    Estimate (total dollar amount) of damage and lost revenues.
    Measured acceleration levels.

When investigating a harbor facility, we are typically interested in obtaining a brief description of the portions of the facility listed below, including the conditions of these facilities (any existing damage or damage caused by past seismic events) prior to the earthquake. Next, we would like to document the damage sustained by these portions of the harbor, the impact of that damage on the users, the methods (both permanent and interim) used to restore service, and obstacles (lack of equipment, personnel or materials, lack of access) to the restoration of services.

Navigation Channel
    Information on changes in Channels
    Breakwaters, Jetties
    Seawalls, Quay Walls, Bulkheads, Revetments
    Dikes
    Other retaining structures

Piers and Wharves
    Support structures (piles, caissons, fill, etc.)
    Deck
    Pier or wharf/shoreline interface
    Utility lines buried in the deck
    Seawalls and other retaining structures

Cargo Handling Equipment
    Container cranes
    Other cranes
    Conveyors
    Stacker and reclaimer equipment
    Chiksans/pipelines
    Ramps
    Specialty equipment

Cargo Storage Areas
    Open lots (asphalt, concrete, unpaved)
    Warehouses
    Tank farms
    Silos
    Container control tower
    Specialty facilities

Inland Transportation Systems
    Roads
    Roadway bridges
    Rail lines
    Railroad bridges
    Pipelines

Interfaces with Other Lifelines
    Power
    Water
    Sewage
    Natural gas
    Communication systems

Other Buildings
    Administration and office buildings
    Fire stations
    Maintenance/repair buildings

Emergency Plans
    Describe the plan (if any).
    How did the plan perform?
    Based on the experiences gained from this earthquake, will changes or improvements be made to the plan?

CHECKLIST FOR POST-EARTHQUAKE INSPECTION OF HIGHWAY ROAD/BRIDGE SYSTEMS		
ELEMENT	CHECKLIST ITEM	POSSIBLE CAUSE(S)
Roadway	Localized cracking, settlement Heaving of adjacent roadway	Settlement, heaving, and/or liquefaction of backfill or underlying soil materials
	Sandboils in adjacent soil materials	Soil liquefaction
	Widespread and deep extensional cracking and/or slumping	Surface fault rupture
	Sloughing/sliding/rockfalls of adjacent hillsides or slopes	Sliding due to liquefaction and/or to strong ground shaking
Bridge Abutments	Rigid body settlement or lateral movement of abutments	Damage to abutment foundation
	Sandboils in adjacent soil materials	Soil liquefaction
	Rigid body tilting of end wall Cracking of abutment walls	Excessive lateral pressure in backfill due to pore pressure buildup or strong ground shaking
	Separation of soil backfill from abutment walls	Damage to abutment foundation
Bridge Column/Pier Supports	Tilting, settlement, horizontal movement of supporting foundation	Damage to pier foundation Movement of subsurface soil materials due to pore pressure buildup or strong ground shaking
	Lateral cracking at connection to bent beam, permanent deformation or rotation at connection to deck	Insufficient strength and/or ductility of beam-pier connection
	Diagonal cracking at pier column connection (at multi-column bent)	Insufficient shear resistance at connection
	Diagonal cracking, spalling, along length of column	Insufficient shear reinforcement
	Transverse cracking of column near top or bottom supports	Insufficient longitudinal reinforcement
Bridge Deck	Sliding, scratchmarks, rubbing, spalling, permanent movement at expansion joint	Longitudinal forces that exceed frictional resistance at joint
	Vertical transverse cracking	Insufficient longitudinal reinforcement
	Diagonal cracking	Insufficient shear reinforcement

10. Telecommunications Systems

    System Configuration
    Facilities
    Operations
    Emergency Response
    What to Look For
    Post Earthquake Investigation Procedures and Guidance
    Telecommunications Check Lists

10. TELECOMMUNICATIONS SYSTEMS

Many engineers and responsible civic leaders have developed their own perceptions of what may happen to telecommunications following a major earthquake. These perceptions usually are illustrated by scenes like the Sylmar central office in the San Fernando earthquake of 1971. That was where the equipment bays toppled over like dominoes and interrupted service for many weeks as the central office was rebuilt. Another common scenario is the loss of overhead lines or underground lines. The belief is that loss of lines will cripple the telecommunications network.

SYSTEM CONFIGURATION

With the modern telecommunications networks of today, the total loss of an office is not very likely. The network is made up of so many parts and systems that loss of one system will not affect all the others. Alternative routes or paths are available to bypass problems. Fiber optic transmission systems have been installed in many areas to provide higher quality and higher capacity transmission. The fiber systems have added another medium to further diversify the already varied transmission capabilities.

Switches in telecommunications offices are normally provided with capabilities to maintain service despite maintenance requirements or failures. Standby processing equipment is available to avoid switch shutdown. Loss of a few frames in the switch would not affect the office.

Power rooms are designed with reliability in mind. Individual power plants are provided for each switch installed in an office. These plants are designed with the capacity to carry the loads even with loss of parts of that plant. Because central office equipment operates on -48 volts DC., battery plants work ideally as an emergency source. Rectifiers keep the batteries charged and provide the D.C. source during everyday operation. Capacity is available to carry the office load until emergency generators are started or line power is restored.

The network can also be looked at as a system relying on a highway of cables that interconnect all the different equipment from one location to another. Thousands of miles of power cable, transmission cables, signaling cables, alarm cables run through the central office. Cableways are provided to support all of these runs and provide some means to control the flow of these cables.

As the network becomes more sophisticated and more services are offered, data equipment is relied upon more to provide these functions. Many offices now appear more like data centers than the traditional central office. Mainframe computers are used as part of the network. The evolution of the central office will make traditional network design concepts out of date.

FACILITIES

The equipment providing telecommunications service today is in fact a computer. Mechanical relays have been replaced with microswitches. Processors control the function of the equipment. The equipment processes inputs and outputs in digital format. Being computers the equipment must now function in an environment suited for data equipment. That means temperatures, humidity, shock and vibration, dust must all be controlled to assure reliability.

The air handling equipment in a central office plays a large role in long term electronics reliability. Understanding this role, most central office facilities have provided appropriate capacity for maintaining this environment even with partial equipment failure. Backup power is normally available to keep the equipment functioning even with loss of line power. The electronics equipment is not so vulnerable as to fail immediately with loss of cooling. Most telecommunications equipment will operate for extended periods without air conditioning.

The AC power coming into the central office is distributed in the power room to distribution panels. The AC panel also ties to transfer switches that allow emergency generators to pick up office loads in case of power failure. The office loads may be transferred in steps by automated switches. Central office equipment has priority on power restoration. Building loads would be loaded after the communications equipment is assured of restored. For powering the communications equipment, the AC is rectified to DC voltage. DC distribution panels are located in various areas of the central office to supply the equipment. Normally DC panels are dedicated for each type of equipment.

Buildings have been designed to carry the weight of the equipment. Floors, columns and walls would be stronger than in a people occupied building. The stronger buildings may allow it to tolerate more severe ground motions.

OPERATIONS

The larger central offices are usually manned with personnel most of the day. Small offices may have personnel part of the day. Remote offices require no personnel for operation of the equipment. These remote offices are monitored by alarms and other status equipment by personnel in central locations. The entire network is also monitored by the central control centers for any service problems. Manual intervention is normally not required unless there is a major disruption. Primary tasks of personnel are maintenance functions, responding to trouble reports or service changes.

Most systems have been designed with backup capability in case of problems. If needed, the network is supported with key spare equipment to maintain service. All critical equipment is backed with replacement plug in boards stored in the office. Some equipment requiring disk drives may even have replacement drives stored in the office. The extent of maintaining spare equipment may even include portable transmission systems or switches on trailers. The portable equipment may play double duty in providing services for special events when not used to restore service.

Test equipment is available in each office for troubleshooting or self diagnosis of the network. The diagnosis is a scheduled program that is continually run. Remote equipment is also available for personnel to take to remote locations or other offices.

Power equipment is also available to restore AC power where necessary. Portable generators on trailers are usually placed strategically to move to any location that may need them. Engineers have designed battery plants with enough reserve time to allow movement of these generators into place.

EMERGENCY RESPONSE

Every company has the responsibility to develop their own emergency response programs, 24 hours services are always available. These plans should include personnel protection, restoration procedures, emergency response drills, officer notification, central planning center, and coordinated efforts with other companies and emergency agencies. The plans must be shared with all company personnel and each person must understand their roles in case of disaster.

If the company has an established program in place and is responding to an emergency outside interference should not be tolerated. Disasters normally create enough confusion without having outside agencies attempting to assist with unplanned efforts. The company should have designated emergency response personnel with proper identification allowing their authorized presence.

Customers must understand that in case of service outage, a restoration plan is in place to control demand for service. Local emergency agencies, national security services, key local civic leaders have priority over many other customers. Some industries that rely on communications for economic survival, ie. banks, insurance, aerospace may find that they do not have the highest priority for restoration.

Most phones may experience busy signals immediately following a major disaster. The inability to get dial tone is not necessarily because of equipment failure as must as overload of traffic. How soon this situation resolves itself is dependent on customer activity. Emergency calls may have to be made at pay phones or by going directly to emergency service locations. Pay phones if owned by local telecommunications company are normally classified as critical service allowing access.

WHAT TO LOOK FOR

It may be very difficult to follow the network design of a particular office. Every company may have their own scheme for interconnecting equipment. Grounding schemes may be different, location of equipment may be different and can make the task of surveying an office challenging.

Most personnel working in the central offices may not have full understanding of the equipment layout. Therefore, their input is not always reliable. Personnel working in the toll area may have no idea what goes on in the switch room. Power is only understood by power maintenance people. Even then, the knowledge of the field personnel would be restricted to their day to day function, not the design concepts.

The obvious discrepancies may be the only areas that can be reported. How those discrepancies affect the network will not be ascertained. There are many modes of failures that may not affect the reliability of the network. The recommendation is to report any failed mechanical part, any falling parts, any buckling of structures, loss of service of equipment, failed bracing or anchorages. Poor engineering design in support of equipment should also be noted. Look for any taut cables that may have been stretched by movement of equipment or cableracks.

Evidence of large displacement and stress should be sought. Markings on walls or floors or on equipment panels will give indication of the movement within the office. Buckled structural members may show loads experienced by the equipment.

Common Failures

Expected problems are mechanical failures to overhead ironwork, equipment racks or maybe overturned storage cabinets. Desk-top equipment such as personal computers, typewriters, calculators may end up on the floor. As mentioned earlier, the network is composed of miles of cables interconnecting frames and equipment. The cables are supported by ladder type cableracks that may be severely overloaded in some older and larger office. Depending on the intensity of the ground motion, these cableracks may collapse from failed ceiling anchors or failed rod supports. The experience has been that ironwork hardware will rain onto the floor. The hardware will consist of metal clips, corner brackets, nuts, bolts and splice pieces. Most companies use friction fittings to join the ironwork.

Embedded concrete anchors will most likely be a problem. Tall slender relay racks loaded with equipment shelves are the most common means to mount equipment. These racks are mounted as cantilevers from the floors applying high stress to the anchors when racks are rocked. While top support is provided for the racks, the bracing does not prevent rack top displacement. Anchors may creep or ultimately fail from the severe loads.

Sheetmetal or plastic trim on equipment bays will fall. Most manufacturers use snap on or quick release features to secure panels. They will not withstand much ground motion.

Battery racks that are not provided with adequate cell restraints or not designed for earthquake loads may experience power loss. The cells on the racks may be thrown around breaking intercell connectors or posts or at worst fall off the rack. Jars could break spilling electrolyte if they impact rack uprights or adjoining cells.

Circuit board walkout is a very real problem. Older equipment were not provided with latches to prevent cards from coming. loose. The clamping force of the electrical connector in most cases is not enough to prevent card walkout. A circuit board decoupled from the connector would loose service. Critical cards could shut down the system.

The distributing frames in most offices are large structures made of channels and angles. Channels cantilever out from a center structure that is anchored and overhead braced. At the ends of the channels are terminal blocks with wiring pins. Cable is also supported by the arms of these channels. A lot of mass is supported by these frames with very little bracing to restrain displacement. Damage is expected to the frame, however, service should not be affected severely.

The overall expectation is that there will be failures of isolated equipment and some mechanical damage. Service is not going to be totally affected, however. Limited loss of service may affect some customers. Mechanical restoration of operating equipment may be necessary following the event. The immediate activity would be concentrated on restoration of service where affected, however. Limited loss of service may affect some customers. Mechanical restoration of operating equipment may be necessary following the event. The immediate activity would be concentrated on restoration of service where affected. Many of the concerned failure modes such as downed lines or cut lines may turn out to be ill founded because of the protection built into the network.

POST-EARTHQUAKE INVESTIGATION PROCEDURES AND GUIDANCE

The inspection procedures that follow assumes investigation personnel have some familiarity with central office equipment. The inspection process is usually a visual one, while in some cases cover removal is required for the inspection, this should only be done by telephone company personnel. The inspection of anchor tightness may involve the use of tools to check for nut or bolt torque, however, this must be done by telephone company personnel.

The earthquake investigator should not touch equipment much less attempt corrective actions of any kind. Checking should be limited to visual inspection. Circuit boards that may appear to be loose in shelves or cardcages could be bad boards or incorrect boards that were stored there prior to the earthquake. Fuses may be pulled purposely prior to the earthquake.

Due to densely packaged integrated circuits in newer equipment, inspection personnel should not touch the cabinets or frames to avoid ESD (electrostatic discharge) which could cause damage to circuitry.

It is recommended that earthquake investigator be accompanied by people that are familiar with that particular office or with the equipment. The Equipment Engineers, COE's, DEC Engineer or Maintenance Engineer would be good resources. It is a good practice to report to the staff accompanying you your findings before leaving the building.

Cable racks run through the central office as continuous systems. Though the inspection procedures are divided into specific equipment areas, the cable rack inspections must be followed through other equipment areas, and through wall or floor openings.

Power Room

Extra care must be exercised to avoid injury. Don't touch any object if you do not thoroughly understand its operation and function.

- Examine the battery rack verticals for buckling or bends. Fresh paint crack lines are good indication of bending.

- Inspect battery rack side and end cell rails for failure, bends or looseness.
- Examine battery rack joints for weld failures or loose fasteners. Weld failures could be hairline fractures so the use of magnifying glass may be needed.
- Inspect battery cell tray or rail supports for bends or buckling as well as acid spill. Inform engineer(s) when acid spill is found.
- If there is overhead braced, check for bends, loose or failure of ironwork.
- Examine anchors for looseness.
- Examine cell jar for cracks or separation of top from jar. Cracks could be hairline in size. Look especially at areas where jar may have impacted rack and at bottom of jar.
- Examine cell post seal area for bent or broken posts. Inspect for bent, loose or shorted intercell straps on batteries. Look for cell posts that may have been broken by intercell straps. Look into jar, which are usually transparent, and check for post separation from plates.
- Inspect plates inside cell jar for broken plates, dislocated, and mis-aligned plate supports. Plates should be evenly supported across the jar.
- Inspect inter-row and inter-tier cables for looseness or bent post connector plates. Cables to busbar may be separated or be chafed.
- Verify ground cable to rack is secure.

Rectifiers and Distribution Panels

- Inspect for operation by looking at output meters. Some rectifiers may not be carrying load with output meters reading zero.
- Inspect cabinet condition for open doors, bent or buckled cabinet panels, bent structural frame, and loose equipment within panel. Check transformer mounts for damage or looseness.
- Visually inspect cable connections for tightness, broken cables, and damaged cable insulation. Review AC input cables, DC output cables and ground connections.
- Inspect for busbar connection tightness and slippage. Check for possible shorts or opens.
- Inspect cabinet anchors and/or overhead bracing.
- Check for loose circuit boards.
- Check for meter operation and loose panel instruments. Check for loose fuses or broken fuse holders. Ask craftsperson if circuit breakers have to be reset and how many.

Busbar

- Inspect all joints for bolts for tightness and slippage. Follow busbar from one end to the other.

- Check all insulators for cracks, breakage or misalignment.

- Look for ironwork, electrical conduit, water pipes, lighting fixtures, ventilation ducts that have fallen over or near busbar or battery racks.

Engine-Generators

- Inspect all fuel, water and air lines for broken joints and leaks. Bent pipes may exist at floor or wall openings. Pipe integrity must be checked along length of pipe.

- Intake air and exhaust piping should be inspected for collapse or bends. Flexible joints may be crushed or loose. Check integrity of pipe supports.

- Inspect outside radiator for structural damage. Check all legs of radiator for bends or buckling. Check fan motor mounts for tears, looseness or failure. Check heat exchanger tubes and header for coolant leaks. Inspect anchors for looseness or failure.

- Inspect engine skid mounts and anchors for damage. Skids may have isolators that are steel springs or rubber. Check for broken springs or metal parts in isolators. Inspect anchors for looseness or broken concrete. Check snubber for signs of engine displacement.

- Check generator housing for mechanical problems. Inspect bearing supports for damage. Check coupling to engine for damage. Verify cables to generator are intact and not shorted.

- Inspect starting battery racks and cells. Inspect charger for operation. Wall mounted chargers may have come loose.

- Inspect air storage tanks and compressors for anchorage and leakage. All pipe joints must be checked for looseness. Check pressure gauges for pressure.

- Enquire about start up problems.

Main Distributing Frame (MDF) and Distribution Frames

- Inspect distributing frame for displacement and out of square. Check for sheared frame bolts.

- Check for bent or buckled uprights, horizontals for block mounting strips.

- Inspect anchors for looseness and bent base angles.

- Verify ground strap is intact with ground bar. Check cable rack and Auxiliary Framing (overhead ironwork) above for signs of damage such as loosen friction clips.

- Inspect cable rack above frame for failed mounts or joints.
- Check wiring to blocks for broken leads or tautness of wires.
- Check for loose or fallen connector blocks, protectors, and protector mounts.

Cable Vault

CAUTION: Inspections in the cable vault should be conducted only after the vault has been checked for toxic and flammable gases by qualified personnel, and should always be accompanied by qualified/personal or person responsible for the cablevault.

- Verify that cable entry wall seals are still good. Check for cracked sealing material or cracked concrete around entry openings.
- Check for signs of cable pull or chafing of cable insulation.
- Check cable splice cases that may have fallen off rack. Splice cases may have cracks or opened.
- Check support structure for wall anchor condition, loose fasteners, or bent ironwork.
- Visually check for operation of pressurizing equipment by looking for pressure at gauges.
- Inspect pressurization equipment cabinet condition for bent panels or loose parts. Inspect anchorage of cabinets for looseness.
- Inspect pressure tubing connections for pulled or broken joints.

Switching Equipment

- Inspect equipment frame for buckled uprights, shifted baseplate, loose doors, covers and unsupported equipment shelves. Examine card cages for distorted frames and cracked back-plane and connectors.
- Check for loose side panel covers, cable tray panels, cross aisle cable trays, frame joining hardware.
- Examine anchor for looseness or broken concrete. Some frames may have hidden anchors requiring opening of covers. Random inspection is required where there are obvious signs of equipment frame movement. The movement can be traced by auxiliary framing hitting walls, columns or buckled framing. Fresh crack paint lines can be a clue.
- Check overhead bracing, if used, for buckled or slipped bracing. All fasteners should be tight.
- Check cable strain reliefs.

Equipment Performance

- Inquire with Operations Personnel of any service problems that resulted from the earthquake, such as loose circuit boards, system re-sets, open fuses, loose connectors.

- Check for loose plug in circuit boards in equipment cages and loose wires in backplane. Look for taut cables and wires in cable tray.

- Check for loose fuses or blown fuses in fuse panels.

- For equipment with cooling fans, check for proper fan operation and verify filters have not come out of housing.

- Inspect tape drive or disk drive units for proper operation. Tape drive unit may have lost tape spool. Inquire of operations personnel.

- Check for cooling fans operation.

Overhead Auxiliary Framing for Cableways

- Inspect framing for loose channels, ladder rack or braces. Follow cable racks through wall and floor openings. Look for clips, support brackets, hangers, splices that may have come loose. Heavily loaded cable racks are especially vulnerable to failure.

- Diagonal cable rack or vertical cable rack may have slipped junction hardware (usually friction clips) which will not be an obvious problem. Slipped joints may fail with aftershocks.

- Check for cables that have fallen off cable racks. Cables unsecured and piled over cable horn height could fall off cableracks.

- Check for ceiling anchor failure or hanger clips that have pulled through channels. Hanger failure may have dropped framing and cable rack on top of lower cable racks or equipment.

Switch Frame Mounted Cable Racks

- Inspect sheetmetal cable racks for damage or loose panels. Normally cables are unsecured and overloaded racks may have bowed panels.

- Cross aisle cableracks should be inspected for proper fastening. Sheetmetal screws may be pulled out or not exist. These cross aisle racks may have telescoping sections that could fail.

- Inspect cables for chafed insulation from sheetmetal edges.

Lighting and AC Distribution

- Inspect lighting fixtures for slipped hangers or failed hangers. Conduit may have pulled out of fixtures or junction boxes.

- Check for broken tubes and tube restraints in fixtures.
- Check for alignment of light fixtures.
- Inspect all AC conduit for separation from junction boxes. Separation of solid conduit from junction box occurs frequently because of out of phase motions of equipment and building walls.

Grounding

- Check ground cables to ground bar connections are intact and fasteners are tight.
- Check power ditribution panel for shorted cables or loose connections.

Other Switch Room Equipment

- Check for desktop equipment, such as monitors, keyboard, phone set, etc. that may have fallen or moved.
- Data cabinets may be provided to house datasets for special services. Verify anchorage and bracing of cabinets. Check for damage to datasets or disk drives in cabinets.
- Check spare parts storage cabinets for damage or toppling. Verify anchorages had been provided.
- Check spares storage cabinets or shelves for damage and conditions of spare parts

Building Condition - Environmental Problems

- Inspect for water leaks over equipment frames.
- Verify cooling equipment is functioning and air ducts have not fallen on equipment.
- Check for concrete dust or debris on equipment that may contaminate electrical contacts.

Fiber Optic Cable

- Check fiber for tension or crimps. The fiber may have been pulled taut by collapsed cablerack or movement of equipment. No tension is allowed on fiber. Bends or crimps in the fiber will affect service. Bend radii of fibre is normally between 2 to 4 inches.
- Fiber trays or cablerack condition should be checked for adequate support. These trays may be mounted to building walls or mounted to auxiliary framing using nonstandard hardware.

Microwave Tower

- Inspect tower structure for damage. Look for bent or buckled ironwork. Check for loose bolts, braces, or joints.

- Inspect antenna for loose or broken mounts. A telescope is helpful.

- Look for waveguide damage and proper support. Rigid waveguide should not be kinked or buckled. There may be rigid bends or curves formed into some of the waveguide. Corrugated flexible waveguide joints should be inspected for looseness.

Microwave Radio Equipment

- Inspect radio for damage at rack mounts or cabinet mounts.

- Inspect waveguide connections to radio for tightness. Look for kinks or bends of rigid waveguide. Corrugated flexible waveguide joints should be inspected for looseness.

Data Processing Equipment Room

Use same procedures as for Switching Equipment Room with these additional inspections.

- Inspect equipment for damage to frame or panels. If equipment has overturned or fallen into cable openings, inspect for structural damage.

- Check for falling equipment on cabinets.

- Data cabinets that have moved or toppled may have damaged electronics. Disk drives or tape drives may have been damaged. Inspect for obvious physical damage.

- Check equipment for loose cable connectors or broken cables.

- Check for water under floor panels. Water could get into AC junction boxes and short across leads.

- Check for cooling fan operation in equipment so equipped.

- Inspect raised floor pedestals for collapse or looseness. The pedestals may have broken base welds, bent tubes, bent base or failed anchors.

- Inspect raised floor stringers between pedestals for buckling or tearing at pedestals.

- Inspect raised floor panels for collapse or pop out from stringers. Floor panels could pop out when stringers are compressed against walls.

- Verify flooring grounding straps have not broken.

Uninterrupted Power Supply (UPS)

- Inspect UPS distribution panels for cabinet damage.
- Check cable connections for tightness. Verify grounding cables are intact.
- Inspect UPS battery racks for damage using the Power Room section for details.
- Inspect cable racks for support and joint integrity.

Air Conditioner

- Inspect air conditioner housing for damage. Anchorage of air conditioner should be checked for tightness.
- Check for water leaks at inlet and outlet pipe connections at room unit.
- Check radiator condition outside of building. Water pipe connections need to be checked for leaks.
- Verify operation by noting temperature and humidity.
- Check for water tank damage.

TELECOMMUNICATIONS CHECK LIST

Power Room

Batteries
__ Battery rack
__ Cell jar integrity
__ Intercell connectors
__ Anchor condition
__ Battery rack condition

Rectifiers and Distribution Panels
__ Cabinet condition
__ Anchor condition
__ Power cable connections
__ Input/Output

Busbar
__ Cable connections
__ Splice joints
__ Buckling or short
__ Mechanical support

Engine Generators
__ Fuel and water lines
__ Engine skid mounts
__ Intake air and exhaust pipes
__ Generator housing
__ Start battery rack
__ Start air tanks and compressors
__ Oil or water leaks
__ Power output cables
__ Outside radiator

Distribution Frame

Frame Structure
__ Mechanical integrity
__ Anchor condition
__ Overhead supports

Terminal Blocks
__ Wiring condition
__ Block mounting
__ Protector capsules

Switch Room

Switch Equipment
__ Functional capability
__ Cabinet integrity
__ Anchor condition
__ Circuit board walkout
__ Alarm condition
__ Back cabling

Overhead Ironwork
__ Mechanical integrity
__ Falling hardware
__ Cable condition
__ Ceiling anchor condition

Switch Room Operating Area
__ Desk-top equipment condition
__ Data cabinet condition
__ Spare parts storage

Transmission Room

Transmission Equipment
__ Functional capability
__ Mounting rack integrity
__ Anchor condition
__ Circuit board walkout
__ Alarm condition
__ Overhead bracing condition
__ Back cabling

Overhead Ironwork
__ Mechanical integrity
__ Falling hardware
__ Cable condition
__ Ceiling anchor condition

Transmission room control area
__ Desktop equipment
__ Spare parts storage

Building Facilities

Structures
__ Ceiling condition
__ Floor condition
__ Staircases/Stairwells
__ Joints
__ Inside and outside walls

Elevators
__ Operating status
__ Landing condition

Heating, Ventilation and Air Conditioning
__ Operating status
__ Mechanical integrity
__ Pipe hangers
__ Heat exchanger
__ Coolant leaks
__ Duct condition
__ Power panel

Data Room

Data Equipment
__ Equipment operation
__ Cabinet integrity
__ Cable condition
__ Disk drive condition
__ Tape drive condition

Raised Floor
__ Pedestal condition
__ Floor panel condition

Power Distribution
__ Distribution panel condition
__ UPS battery condition

Air Conditioning
__ Cabinet integrity
__ Water pipes
__ Cabinet bracing and anchors
__ Operation

Cable Vault

Cable Supports
__ Cable condition
__ Wall anchors
__ Outer wall opening condition

Pressurization Equipment
__ Operation
__ Cabinet condition
__ Anchor condition
__ Tube connections

11. Liquefied Fuels and Natural Gas Systems

   System Configurations
   Local Natural Gas Distribution Companies
   System Components: Seismic Performance
   Primary Issues for Post-Earthquake Investigation
   Some Questions to be Addressed By Reconnaissance
      Findings
   Gas and Liquid Fuel Lifeline Issues Check Lists
   Pipeline Failures Check List
   Customer Related Equipment Check List
   Operations and Maintenance Check List

11. NATURAL GAS AND LIQUID FUELS PIPELINE SYSTEMS

SYSTEM CONFIGURATIONS

Natural gas and liquid fuels pipeline systems are vital components of residential and commercial energy distribution systems and manufacturing support services. Liquid fuels is a generic term that includes not only crude oil but refined products such as gasoline, jet fuel, liquefied petroleum gas (LPG), lubrication oil and diesel fuel. Main components of the pipeline systems used to transport these materials are quite similar. Differences are primarily related to the process of piping gas versus liquids, ties to a local distribution system for natural gas, and the variety of storage devices used for natural gas (e.g., gas fields, gas holders, liquefied gas tanks).

Crude oil and natural gas are often taken from the same field. Product from the wellheads is routed through a system of gathering lines. These lines may be buried or may simply lay on the ground. Gas generally exits the wellhead under a pressure of several hundred psi. Hydration equipment may be required to remove liquids from the gas stream before it proceeds to a manifold for metering of volume and energy content. The gas is then ready to enter a transmission pipeline. Oil is often extracted with the assistance of a large walking beam pump. Gathering lines may converge at a temporary storage facility where additional pumping stations may be needed to send oil through a transmission line.

Transmission lines for natural gas are indistinguishable from oil transmission lines. The transmission lines are typically constructed of high-strength carbon steel and, with very few exceptions, butt-welded joints. Pipe sizes range from 8 inches to over 30 inches in diameter with typical operating pressures of 125 psi to 1000 psi. Corrosion protection generally includes a protective wrapping or coating, and cathodic protection with sacrificial anodes or impressed DC current. Above ground pumping (oil) and compressor (gas) stations are placed along the transmission pipeline to ensure the required pressure head needed to move the product. Other major above ground facilities include metering stations at locations where product may be transferred along the transmission route. In addition, many block valves may be located along the transmission pipeline to provide isolation of portions of the pipeline. These block valves may be remotely operated and typically have a small surface structure to allow access the to valves for maintenance.

It is common practice in the United States to monitor and control the flow of product in transmission pipelines electronically around the clock through a central office. Information from remote sites is telemetered to control centers over telephone lines or through the use of a company microwave system. Remote valves, compressors, pumps, and pressure regulators can be operated from the control center by the same process. Irregularities in the line pressure, product flow rate or other monitored condition are evaluated as they occur. Control center personnel are trained to respond rapidly in emergency situation in accordance with standard operating procedures and detailed emergency plans.

It should be recognized that this practice is not common in some lesser developed countries. Control of the transmission pipeline may be contingent upon maintaining communication between the terminal points of the pipeline and the ability to notify personnel along the pipeline to close manual block valves. Modes of communication include land lines, microwave, and radio.

Oil is eventually delivered to a central storage terminal. This may be part of a refinery where the lighter petroleum products are produced. The storage terminal may simply serve as a holding point for oil to be transferred to tankers, barges, or other transmission pipelines. Refineries are the beginning points for other pipelines that distribute refined product to customers or other distribution centers. Oil products are typically stored in separate tanks within the refinery. Plant piping and manifolds allow a single pipeline to be used to ship a variety of liquid products such as gasoline, diesel fuel and LPG.

LOCAL NATURAL GAS DISTRIBUTION COMPANIES

Natural gas is delivered to local distribution companies (LDC's) via a metering station and associated distribution piping located along the transmission pipeline. At this point, the LDC piping is similar to the transmission pipeline as a result of the elevated pressures involved.

After gas enters the LDC's network, pressures are reduced at facilities variously identified as "pressure regulator stations" or "governors". There, pressures are lowered to one-quarter psig through 60 psig, depending on downstream piping materials and company practice.

Within an urban area it is not unusual to find underground gas piping in most, if not all streets. Piping is typically interconnected in a grid pattern with the largest diameter pipes immediately downstream of the pressure regulator station. Smaller diameter pipes branch off of these at street intersections, with even smaller pipes extending into cul-de-sacs or dead ends.

The proportional size of these pipes is purely dependent on the gas pressure involved. For example, on a low pressure (0.25 psig) system the piping immediately downstream of the pressure regulator may commonly be 12-inch diameter, with the majority of the branch piping being 6-inch diameter and the smallest pipe diameter being 4 inches. These relatively large diameters are required to limit the pressure drop (approximately 0.1 psig is required at the burner tip).

Systems that operate at higher pressure might have 6 or even 4-inch piping coming directly from the pressure regulator station, with 2-inch being the predominant size of branching main. Piping of 1-1/4 inches would probably be the smallest diameter main encountered.

Most LDC's also have numerous valves in their systems, assessed through manholes or small valve boxes. These are usually configured to enable the company to isolate segments of the system for emergency shutdown purposes.

While it is common for gas pressure and flow to be electronically monitored and controlled through a central office within a transmission pipeline system, this is much less common, though not unknown, in a distribution system. Because of the innate complexity of a typical urban gas distribution grid and the large number of pressure regulator station involved, pressures are usually set manually at the station, with scheduled inspections made to monitor conditions.

When compared to transmission pressure piping, LDC piping may vary widely. Besides wrapped, weld steel, one might encounter bare steel, compression-coupled joints, polyethylene plastic (PE) and cast iron with leaded joints. Cast iron or bare steel pipe over 30 years old might still be in use and providing satisfactory service.

"Service lines", the small-diameter pipes that bring the gas from the street mains to the customer's piping, also may be constructed of wrapped or bare steel, PE or wrought iron (possibly with threaded joints). In addition, older service lines may have been inserted with smaller diameter PE or copper.

The LDC's piping terminates with a meter and possibly pressure reduction equipment at the customer's property line or building wall. Downstream of this, the customer's appliances or gas-fired equipment are located. Interior gas piping is almost exclusively made of bare steel, occasionally with welded, but usually threaded joints. In the future, the use of thin-wall, flexible tubing may occur. Flexible tubing is sometimes used at the residential customer's appliance.

Other significant but relatively uncommon components of an LDC's system may be large, low pressure gas storage "holders", most over forty years old; above ground, horizontal propane tanks used for "peak shaving"; and large, liquified natural gas tanks used for the same purpose.

SYSTEM COMPONENTS: SEISMIC PERFORMANCE

The investigation of the seismic performance of natural gas facilities will inevitably require significant interaction between the investigator and the system owner/operator. The scope of damage to the system will rarely be obvious to the investigator. Also, damage to the system that result in product loss will likely be repaired prior to the investigation. Some of the typical earthquake-induced failures that are common to oil and gas systems are discussed below.

Piping

In the absence of corrosion, it is rare that welded steel pipelines will exhibit damage as a result of ground shaking. This is primarily a result of the small ground strains that occur in all but the weakest soil deposits. Exceptions may be found at transitions between vastly different soil deposits, connections to rigid structures, and branch connections to other piping. These exceptions are normally limited to natural gas LDC piping, refinery facilities, and product handling terminals.

Permanent ground deformation is the most severe earthquake-related condition affecting buried pipelines. Surface faulting is an obvious example. Other sources of permanent ground deformation include lateral spreading, liquefaction related settlement, and earthquake-activated landslides.

Continuous welded steel pipelines may be designed to withstand several feet of permanent ground deformation. Older pipelines, which have no special seismic design considerations, may have much less capacity. Other factors may reduce the resistance of older pipelines to permanent ground deformation. Pipelines constructed prior to the 1930's often used oxyacetylene welding which produces weaker and more brittle welds than the arc welding process which followed it. Some older pipe may also have no special corrosion protection and may not have the capacity to undergo additional loading imposed by permanent ground deformation.

Compared to welded steel pipelines, most LDC piping has relatively little resistance to permanent ground deformation. Threaded pipe connections are subject to stress related fracture at the threads. Jointed cast iron or ductile iron are susceptible to joint pull-out. In addition, cast iron can experience brittle failure within the body of the pipe at relatively small bending stresses. Burial and backfill requirements of LDC piping, which are often located under streets in urbanized areas, are often quite different from transmission pipelines. Burial depths may vary greatly over short distances to accommodate other utilities and backfill is often required to be firmly compacted to reduce the potential for street settlement.

Since 1970, polyethylene (PE) plastic pipe has become widely used in new gas distribution systems. Smaller diameter PE pipe is also commonly inserted into low pressure jointed cast iron and steel pipe as part of a system upgrade. The smaller pipe can maintain the level of gas service by being operated at a higher pressure. Earthquake performance of PE pipe, either plain or inserted into existing pipe, has very limited earthquake experience. There were no failures reported in the Loma Prieta earthquake. Significant ductility of the pipe and the connections combined with the low friction resistance between the pipe and soil are factors that indicate that PE pipe is less sensitive to ground shaking effects. The resistance of PE pipe to permanent ground deformations is certainly less than welded steel but may be greater than the jointed pipe presently composing a majority of LDC systems.

Special Pipeline Structures

Transmission pipelines traverse a wide variety of terrain. Special construction techniques and/or structures are often used to cross highways, rivers, ravines, flood plains and wetland areas. Oftentimes, these crossings are above ground and may involve substantial support structures. The effects of an earthquake on these special structures, which may not have been accounted for, often exacerbates the problems faced in the original design.

Wellheads

Because of their small size and component mass, it is doubtful that valve manifolds at wellheads will experience any failures aside from minor flange leakage during significant seismic shaking.

Compressor and Pumping Stations

Possible damage include pipe failure due to the movement of equipment and flange leakage. Loss of telemetered control due to electronic equipment movement, loss of power or misalignment of microwave dishes are possible. If the equipment is housed in brick or masonry building, structure failure may cause equipment failure. Compressor and pumping stations may have emergency power to maintain telemetered control of the pipeline. Depending upon the material being transported, loss of electrical power may trigger an automatic system reaction. This is more common in gas transmission than with liquid fuels, which may require more continuous control to avoid pressure transients.

Transmission Metering Stations

Loss of telemetered data or building failure are the most likely damages to be expected. Minor flange leaks may occur.

Pressure Regulator Stations

These facilities may be housed in underground vaults, structures or simply within fenced enclosures. Information is not typically telemetered off site. Expected damage might include flange leaks or failures of small diameter pressure-sensing lines due to building structure collapse. However, damage to this equipment may lead to over pressurizing the downstream piping system, with possible catastrophic effect.

Low Pressure Gas Holders

Old, low pressure gas holders may fail catastrophically, with a substantial loss of natural gas. This may be caused by misalignment of pressure seals, loss of seals on telescoping water seal holders and deformation of structural members contributing to collapse. Flange leaks at compressors, pressure regulator or vales may occur. Damages expected with older piping (i.e, cast iron, oxyacetylene welded steel) may be present adjacent to the holder.

Underground Natural Gas Storage

Where geologic conditions permit, natural gas may be stored below ground, at high pressure. While this may be accomplished through the use of natural caverns, it is more common for gas to be reinjected into depleted gas fields. These fields consist of permeable rock, rather than caverns or large voids in the rock. It is highly unlikely that seismic activity would disturb these geologic formations to the extent that gas would be lost to the atmosphere. Above ground, man-made equipment failure would pose a much higher probability of failure.

Customer Metering

Above-ground metering equipment, located adjacent to buildings, may experience damage due to component twisting or the collapse of masonry. Threaded or flanged joints may leak. Aluminum parts may be cracked or deformed.

Customer Equipment

All gas equipment, if not properly anchored, is susceptible to sliding or tipping. Shearing of pipes at the equipment interface may occur. Movement or toppling of water heaters at Modified Mercalli Intensity VII can be expected. These types of damage may be independent of the age of the structure. Movement of buildings off their foundations will impact all utility piping into the building.

Of considerable concern to the utility companies is possible customer overreaction to perceived risk of gas leakage after an earthquake. Many more customers may physically turn off the gas to the buildings than need to. After one recent California earthquake it was found that less than 25% of residences in severely damaged areas, where occupants had turned off the gas, actually had leaks. In the Loma Prieta earthquake, about 150,000 services were shut off, and since most customers will call the utility to turn the gas back on, the time required to "re-light" all customers may be greatly extended.

Propane Tanks

In some areas, propane is used for "peak shaving", or maintaining adequate system pressure during times of peak gas usage. Large, above ground propane tanks may be susceptible to sliding, if unanchored, pedestal or support buckling and shearing of piping above ground level. (See Section 12 on tanks).

Liquefied Natural Gas

Liquified natural gas (LNG) tanks and their associated plants are, arguable, the most comprehensively designed feature of a modern natural gas system. Unpressurized, ultra-cold (-260° F) liquid natural gas is contained in these structures. The design, construction and ongoing operation may be closely monitored by state utility commissions, local building officials, and fire departments. It is thus reasonable to assume that seismic considerations, which are required under federal code, were used in the design and more important, emergency response measure are in place to address catastrophic situations. Critical elements may include foundation/soil interaction, shear stiffness of the tank, and the effects of fluid sloshing within the tank. (See Section 12 on tanks.)

PRIMARY ISSUES FOR POST-EARTHQUAKE INVESTIGATION

A check list of key issues pertaining to a post-earthquake reconnaissance of the performance of gas and liquid fuel pipeline systems is below. Not all of the check list items are always applicable or essential as part of the reconnaissance activity. In addition, there is a list of current questions that continued post-earthquake reconnaissance may hopefully answer.

It is almost certain that pipeline system operators will be operating in a mode of heightened activity during reconnaissance activities. Post-earthquake investigation is not a high-priority issue in the days or weeks following a major damaging earthquake. Establishing contacts for future communication within the pipeline system operating company is one of the most important functions of the reconnaissance effort. Unless the damage is visible at the time, all of the information in the check list will be obtained from company records or the recollection of company personnel when the level of company activity permits it.

Information on customer equipment is important for identifying means to increase public safety in future earthquakes. For LDC's, much of the damage to customer equipment is unknown to the company unless observations are documented during service restoration. Also, legal ramifications often prevent any timely disclosure of details on damage related to fire or explosion if it resulted in injury or property damage. Collection of customer equipment information will typically involve a house-to-house survey in a selected area.

SOME QUESTIONS TO BE ADDRESSED BY RECONNAISSANCE FINDINGS

1. Do welded steel pipelines exhibit a relatively uniform deformation pattern prior to tensile failure?
2. Does buried pipeline response correspond to surface evidence of permanent ground deformation?
3. How does the performance of plain or inserted polyethylene pipe compare to other distribution pipelines?
4. For jointed pipelines, are ground motions accommodated by all joints or only a few?
5. What impact do service connections have on the ability of gas distribution pipelines to withstand permanent ground deformation?
6. How significant are the urban piping installation requirements (e.g., burial depth, compaction) to the ability of the buried pipeline to accommodate permanent ground deformation?
7. What impact do automatic gas shut-off valves have in preventing additional damage and the rate of service restoration?
8. What is the relative performance level of different gas appliance connection details?
9. Is emergency preparedness sufficient to restore service when customers are ready to receive service?

GAS AND LIQUID FUEL LIFELINE ISSUES CHECK LISTS

Vulnerability
__ Impact of lifeline failure on public
__ Repair and restoration costs
__ Restoration time
__ Economic impact to customers

Regulatory Policy
__ Appropriate level of protection
__ Regulatory actions
__ Funding
__ Standards

Earthquake Awareness
__ Lesson learned
__ Design/operations guidance

Design
__ Pipeline performance at high strain
__ Soil-pipeline interaction characteristics

PIPELINE FAILURES CHECK LIST

If possible, all piping information should be recorded by individual pipe size and material.

General Conditions
__ Miles of pipe in impacted areas
__ Operating pressure
__ Maximum allowable pressure
__ Size (diameter)
__ Wall thickness
__ Material
__ Pipe grade or specification

Type of Failure
__ Shear   __ Compression
__ Crack   __ Joint pull-out

Location of Failure
__ Pipe   __ Weld or fusion
__ Coupling   __ At bends, tees, valves, etc.

Failure Consequences
__ Fire   __ Explosion
__ System shutdown or underpressure

Probable Cause of Failure
__ Liquefaction   __ Faulting
__ Ground shaking   __ Landslide

Condition of Pipeline
__ Corrosion __ Type of weld or connection

Soil Conditions
__ Soil Type __ Depth to pipe
__ Ground Water __ Special bedding

FACILITY

Type
__ Pump station __ compressor station
__ Metering station __ Pressure regulator station
__ Product storage

Location
__ Above ground __ Vault
__ Building (material)

Interruption of Operation
__ External causes __ Component failure

Facility Performance
__ Equipment __ Piping
__ Structure __ Instrumentation and control
__ Tankage

Failure Consequences
__ Fire __ Explosion
__ System overpressure __ System shutdown or underpressure

CUSTOMER RELATED EQUIPMENT CHECK LIST

Type of Equipment
__ Metering __ Domestic appliances
__ Commercial equipment __ Industrial Equipment

Type of Damage
__ Piping __ Inadequate equipment anchorage

Failure Consequences
__ Fire __ Explosion
__ System overpressure __ System shutdown or underpressure

Emergency Devices
__ Manual valves __ Automatic shut-off valves
__ Did customer know how to shut down system?

OPERATIONS AND MAINTENANCE CHECK LIST

Service Interruptions
__ Number of customers __ Duration of interruption
__ Reduction of throughput

Maintenance and Repair
__ Number of repair orders   __ Numbers and types of emergencies
__ Total cost or man-hours   __ Availability of spare parts
__ Long term increase in leaks

Earthquake Preparedness Assessment
__ Employee training   __ Emergency response plan
__ Internal communications   __ External liaison
__ Maps and records

12. Common Facilities

    Tanks
        Types of Tanks
        Earthquake Performance of Tanks
    Emergency Power
        Back-up Batteries
        Engine-Generators
        Uninterruptable Power Supplies
    Tank Check Lists
    Emergency Power Check Lists

12. COMMON FACILITIES - TANKS

INTRODUCTION

Tanks are used in a broad range of lifeline facilities, and are used for storage of water, fuel, electrical insulating oil, lubricating oil, water treatment chemicals and many other applications. Tank failures cause problems not only because their contents are lost and not available for use, but also because there can be significant secondary impacts, such as the release of toxic or flammable materials.

TYPES OF TANKS

Four different types of tanks are considered here: flat-bottom tanks, horizontal tanks, vertical tanks on short legs, and elevated tanks.

Flat Bottom Tanks

Flat-bottom tanks are the type most commonly used for the storage of large volumes. Typical size vary from less than 10,000 g to over several million g. The name is somewhat deceptive in that most of these tanks have bottoms that are slightly domed, so that contents can drain to the outer edge to a drain pipe. Three major variations are encountered: unanchored, anchored, and floating-roof. Larger tanks may have internal columns to help support the roof. Figure 12.1 shows a schematic diagram of a flat-bottom tank in which common terms used with tanks are defined. Tanks can be placed on a prepared mat, a ring foundation along the perimeter of the tank, or on a ring foundation that has a small lip that confines the edge of the tank. The tanks can be anchored or unanchored. Anchors are typically long "J" bolts that extend deep into the ring foundation and are attached to the tank using one of many different chair designs.

Horizontal Tanks

Horizontal tanks have a circular cylindrical shape and are usually supported in two cradles with the axis of the cylinder parallel to the ground. They range in size from less that 100 g to over 10,000 g.

Vertical Tanks on Short Legs

These tanks take the form of vertical cylinders and are supported on steel legs that are typically welded to the bottom or sides of the tank. These are typically smaller tanks with a capacity of less than 2000 g.

Elevated Tanks

This type of tank is primarily used in water systems and can take many different forms. In the U.S. they are usually a steel tank supported on three or more diagonally braced legs or on a single circular steel pedestal. Outside the U.S., elevated tanks made of concrete are also common.

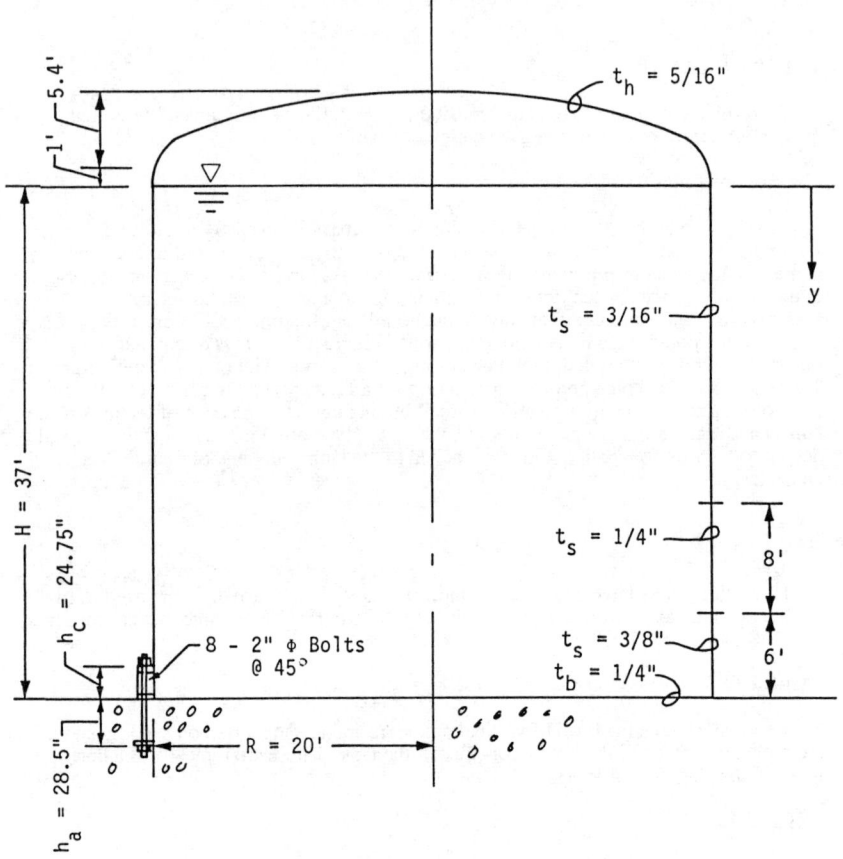

Fig. 12.1 Schematic diagram of flat bottom tank with common terms defined.

EARTHQUAKE PERFORMANCE OF TANKS

Earthquake failures of each of the four types of tanks described above have been observed. Flat bottom tanks are the most vulnerable. While several factors influence tank failure, the height and density of the fluid in the tank at the time of the earthquake is one of the most critical, and unfortunately, one of the more difficult variables to determine. Because flat bottom tanks are the most common and fail most often, emphasis will be placed on this type of tank.

Several failure modes have been observed in flat-bottom tanks. One commonly observes failure, which does not usually lead to the loss of contents is buckling of tank walls. Buckling is usually due to rocking of the tank or differential settlement under the tank. Unanchored tanks with a radius-to-wall thickness ratio of over about 600 have been damaged. Tanks can develop sufficient overturning moment such that the tank will tend to rock and the edge of the tank lifts off of the ground. On the side of the tank opposite the lift there will be large compressive stresses that can cause a bulging at the base of the tank wall, called elephant foot buckling (Fig. 12.2). In general, this will not cause loss of contents. Even in anchored tanks, uplift can cause anchor bolts to stretch or pull out (Fig 12.3), allowing the tank to lift on one side and possibly buckle on the other.

Two other failure modes can occur which usually do cause loss of tank contents: failure of pipe connections, and failure of the weld between the base plate and side wall of the tank. When a tank lifts, if pipe connections are not sufficiently flexible, the pipe will fail allowing the contents to drain. It is important that the first pipe anchor point be sufficiently far from the edge of the tank that the tank can rise without causing failure. The second type of failure that can cause a catastrophic loss of contents is the failure of the weld that attaches the side wall of the tank to its base plate. Uplift or elephant foot buckling, coupled with corrosion-induced weakness and possibly a single-fillet weld between the wall and base plate can open several lineal feet of weld, allowing rapid emptying of the tank (Fig 12.4). This can create a partial vacuum in the tank, causing it to implode (Fig. 12.5). Smaller diameter (less than about 30'), unanchored flat-bottom tanks have been observed to slid on their foundations. Lack of flexibility in pipe connections or drain pipes that extend below the tank can fail. In tanks which are relatively full, the top and upper side walls can buckle due to sloshing of the contents, although this usually does not cause loss of contents. Sloshing of content in floating-roof tanks can damage seals and cause the spilling of content over the top of the side wall.

Liquefaction, lateral spreading, and differential settlement have caused tanks to move, tilt slightly, damage pipe connections, but otherwise have not caused serious damage to the tank.

Problems with horizontal tanks are usually caused by inadequate anchorage - the tank is not positively anchored to its supporting saddles, so that it slips and damages connections. If the saddle is not stiffened, there can be weak-axis bending of the saddle. Typically, the tank is only positively anchored at one saddle so that there can be slipping at the second saddle due to thermal expansion. The saddle to which the tank is anchored must be able to withstand the entire longitudinal load. Figure 12.6 shows a well-anchored horizontal tank supported by saddle stiffened to prevent weak-axis bending.

A common failure in vertical tanks on short legs is buckling of one or more leg. A simple calculation shows that the leg was not designed to take significant lateral seismic loads. Figure 12.7 shows a tank with well braced legs.

Elevated water towers have failed, and stretched or broken tie rods are common.

Fig. 12.2 Example of elephant-foot buckling of unanchored tank. This usually does not cause loss of contents unless the weld between the wall and base fails.

Fig. 12.3 The anchor bolt indicates how high the base of the tank rose from its pad.

Fig. 12.4 A large section of the weld between the tank wall and the base failed, in part due to corrosion.

Fig. 12.5 Failure of base weld allowed rapid emptying of this tank, so that a partial vacuum was created in the tank and it implode.

Fig. 12.6 A well-anchored horizontal tank in which the support saddle has been stiffened to prevent weak-axis bending.

Fig. 12.7 A vertical tanks supported on well braced legs prevents their buckling.

12. COMMON FACILITIES - EMERGENCY POWER

INTRODUCTION

Emergency power supplies are becoming increasingly more common in lifeline systems. They are also used for many industrial applications, and in some states, tall buildings are required to have emergency power units to operate elevators. They are also used in industry for monitoring, control, or communications. Sensitive systems, such as computers, in which a sudden loss of power can result in damage or loss of data, also require special back-up power.

Emergency power usually takes one of three forms: batteries, engine-generators, or an uninterruptable power supply (UPS). Battery back-up systems are usually configured so the batteries normally supply power to the desired equipment, and they are continuously recharged from the commercial source of power. When commercial power is lost, recharging stops, but the battery source of power is not interrupted. Engine-generators with automatic starters usually take about 10 seconds after commercial power is lost before emergency power is restored. In some applications, where a longer power disruption can be tolerated, engine-generators are started manually. UPS systems usually consist of batteries, charger and an inverter integrated into a single enclosure. They are usually dedicated to specific pieces of equipment, such as a computer and its peripherals.

BATTERY BACK-UP SYSTEMS

System Configurations

These systems usually consist of lead-acid batteries supported and restrained by a battery rack. A number of cells are connected in series to obtain the desired voltage. There is a battery charger that operates off of commercial power, and in some cases an engine-generator continues to charge the batteries in the event that commercial power is lost. In some cases, there may also be a converter that converts DC power to AC power to operate equipment requiring AC power. Typically, back-up batteries are housed in their own room along with chargers and inverters.

Earthquake Performance

Seismic performance of battery back-up systems has been mixed. In most cases, battery racks are not designed with earthquakes in mind, so that batteries are supported by the racks, but are not constrained by them. In an earthquake the batteries can fall from the rack and be damaged when they strike the floor (Fig. 12.8). Battery racks designed for earthquakes have side and end rails which serve to restrain the cells to the rack (Fig. 12.9). Even with battery racks designed for earthquakes, there can be problems if the racks are not adequately anchored to the floor, or are not strong enough to withstand earthquake-induced loads. Large cells used by the power and communication industry can weigh over 100 kg each, and a rack may support 15 to 30 cells. Racks typically require diagonal bracing or large support members that can withstand substantial moments. In addition, there should be spacers between cells and between cells and the side and end restraints to prevent impacting during an earthquake.

Fig. 12.8 Typical battery rack without seismic design. Batteries are not secured to the rack.

Fig. 12.9 Seismically-designed battery rack. Batteries are secured to the rack and the rack is braced. The rack must also be anchored.

Without these spacers, cells cases can crack due to impact loads, or there can be internal damage to the cell plates or their supports. Bus bars connecting battery terminal posts or the actual posts have been damaged without spacers. Depending on how the individual cell vent is designed, acid can be spilled, even from a properly mounted cell.

The anchorage of chargers and inverters should also be checked. These units can contain large, heavy transformers. The load path for the transformer anchors and the anchorage of the entire cabinet should be checked for distress. Some units are mounted on structural channels which may be vulnerable to weak-axis bending if they have not been stiffened.

ENGINE-GENERATORS

System Configurations

Engine-generators can range in size from small gasoline-engine powered units weighing less that 100 kg to large diesel units weighing 1000s of kg. Some engines are powered by propane. Most units have a rating of about 500 kVA, are self-contained, and skid-mounted. That is, the control panel, day tank (small local gas supply), and cooling system are all mounted on the skid or attached to the engine-generator directly. Most larger units have the engine-generator mounted on a skid, but the control panel, cooling system, and day tank are independent of the engine-generator. In addition to the small day tank that may hold from 4 to 100 liters, there is usually a large storage tank, generally removed some distance from the engine-generator. Most units are started with batteries similar to an auto-battery. Larger diesel units are typically started with compressed air, so that there may be a small compressor and air tank associated with the emergency generator. For indoor units, provision is made to remove exhaust gases. Larger units are often housed in their own room.

Earthquake Performance

Most engine-generators are relatively rugged because normal operating vibrations can be very severe. However, while the basic engine-generator is earthquake resistant, there are installation practices and design features of auxiliary systems that significantly degrade their earthquake performance.

Engine-generators have experienced a broad range of earthquake-related problems. Engine-generators are typically mounted on skids that are often vibration-isolated to keep engine-generated vibration from being transferred to the the building in which it is located. These mounts are usually composed of springs enclosed in a housing. Many mounts utilize cast iron housings that often fail in an earthquake. As a result, the entire engine-generator can shift 15 cm. or more. This change in position can damage electrical, water, fuel, exhaust system connections and put the unit out of service. More modern installations usually avoid cast iron components and have snubbers that limit the motion of the isolated system (Fig. 12.10). Lack of adequate flexibility and slack in service (fuel, electrical, water, air) connections is a common cause of problems.

Fig. 12.10   Earthquake damaged base-isolation system of the type often used on emergency power engine-generators.

Day tanks are small fuel tanks located near to, or mounted directly on the engine-generator. They are frequently inadequately supported, so that they fail in an earthquake. Fuel lines connecting the main tank to the day tank or connecting the day tank to the engine-generator are often vulnerable to damage, so that they fail and disrupt engine operation.

Auxiliary systems can also cause problems. Vibration-sensitive relays associated with over-temperature or voltage protection, or circuits for the gradual transfer of load to the power units can experience a change-of-state as a result of earthquake-induced vibrations. This can cause a "lock out" situation in which the relays may have to be manually reset. While this can be a relatively simple operation, personnel on duty at the time may not be familiar with the equipment, since it is seldom used, and corrective actions may not be taken. There have been cases where cooling system valves left closed caused the engine-generator to over-heat and shut down shortly after it started. Some diesel engine-generators are normally shut off by closing the air intake duct. There have been cases where this duct was left closed so that the unit would die shortly after it started. The loss of power resulting from poor operating practices can be very disruptive.

There have been several cases in which diesel fuel stored for an emergency generator was degraded with time by growth of organisms in the fuel storage tank. Shortly after systems started, filters or lines become clogged, and the systems shut down. In some cases, the engine-generator would function, but not under full load due to restrictions in the fuel system. This type of problem is typically not found during normal periodic testing, since units are typically tested under light loads. In some cases, additional load is connected to the units over time, so that they are over-loaded and then overheat when pressed into service.

Another common failure of engine-generators is the failure of batteries that are used to start the engine. These batteries are often unsecured, so that they fall over and are damaged in the earthquake and are not available to start the engine. The batteries shown in Figure 12.11, are used to start a diesel that provides emergency power to a communications center at a major airport. Fortunately, the vibrations at the site during the Loma Prieta earthquake did not knock the unrestrained batteries from their support.

Fig. 12.11 Battery used to start an emergency generator that provided power to an airport communications system were unsecured. A half-inch lip on the battery stand provided a little protection.

TANK CHECK LIST - VERTICAL FLAT BOTTOM TANKS

General
__ Facility identification of tank
__ Are detailed tank drawings available? If so attach, and only answer questions not covered by drawings
__ Type of liquid in tank - density and viscosity

Foundation
__ Describe general site conditions - settlement, slope instability, liquefaction (See Site Check List.).

Geometry
__ Tank Diameter
__ Height to top of wall
__ Height of overflow outlet
__ Estimate fluid height at time of earthquake
__ Type of top, Flat, Domed, Floating
__ Material - carbon steel, aluminum, stainless steel
__ Type of wall construction - welded, riveted
__ Wall thickness near base
__ Wall thickness above base
__ Existence of ring stiffeners and location (Clues to internal stiffeners?)
__ Type of support - ground, ring skirt, legs
__ Thickness of scetch plate
__ Type of weld at scetch plate - single or double fillet
__ Type of foundation - soil, asphalt, concrete
__ Condition of foundation - flat, domed, cracks, subsidence
__ Type of anchorage - bolts, straps, retaining ring
__ Spacing of anchorage (on centers) or number of anchors
__ Detailed description of anchor
  Strap width and thickness
  Bolt diameter, length above grade, total length, chair detail
__ Piping connections - diameter, distance to anchor, valves
__ Type of drain - above grade, below grade
__ Vents, type and diameter
__ Height of overflow outlet
__ Height of liquid in tank at time of earthquake
__ Type of liquid in tank - density and viscosity

Signs of Distress
__ Signs of lateral motion - size, direction
__ Failure of drain pipe extending below tank bottom due to tank motion
__ Failure at piping connection due to lateral motion of tank
__ Signs of base lift - stretched, bulled, or broken bolts distressed straps, disturbed ground
__ Elephant foot - height (top to bottom), position of bottom to base, increase in radius, extent around tank, distress of bolts at opposite side, orientation of buck relative to magnetic North, condition of weld
__ Base plate weld failure - length, indications of corrosion, quality of weld
__ Signs of wall distress - chipped paint, leaks, buckles, location
__ Failure of pipe connections

___ Buckling at top of tank - sloshing, implosion
___ Buckling of roof
___ Buckling at stiffener near top of tank
___ Signs of leaks or spills

TANK CHECK LIST - HORIZONTAL TANKS

___ Tank diameter and length
___ Material - carbon steel, aluminum, stainless steel
___ Thickness
___ Shape of tank ends
___ Supported on saddles or legs - number of supports
___ Details of supports - type of material, dimensions
___ Provisions for longitudinal stiffness of supports
___ Details of attachment of tank to supports
___ Indications of tank stiffeners near supports
___ Anchorage of supports to foundation - See Anchorage Check List
___ Details of provision for longitudinal expansion of tank
___ Type of foundation
___ Pipe connections - diameter, flexibility
___ Fullness of tank at time of earthquake, type of material
___ Review Site Check List

Signs of Distress

___ Motion of tank in saddle
___ Distress in supports
___ Distress at support anchorage
___ Action at slip connections
___ Signs of distress - scraped paint, chipped paint, deformation of members

Tanks
___ Description
___ Size and capacity
___ Determine water level at time of event
___ Operational status
___ Foundation- cut or fill
___ Shell condition
___ Inlet-outlet piping
___ Roof structure
___ Elevated tank support structure

EMERGENCY POWER - BACK-UP BATTERIES

___ Is battery rack anchored, braced, and can it withstand lateral loads?
___ Are batteries restrained with spacers between cells, end and side restraints?
___ Look for damage to cell case, post, plates, and bus bars.
___ Look at anchorage and condition of chargers and inverters - check internal transformers.

EMERGENCY POWER - ENGINE-GENERATORS

___ Anchorage of engine-generator (E/G) - snubbers
___ Check slack and stiffness of utility lines to E/G - power, fuel, water, air, and control lines.
___ Check exhaust and intake air ducts.
___ Check day tank and its fuel lines.
___ How is fuel supplied to the day tank - is power needed?
___ Check control cabinet and ask about its function.
___ Check security of starting batteries and connections (for battery start units).
___ Check motor-compressor, isolation, lines, and tank (for air start units).
___ If the unit feeds a unit substation, look at anchorage of dry-type transformer.
___ Ask if unit started and operated as expected.
___ Ask if and how unit is tested - frequency, under load, and how is it test started.

EMERGENCY POWER - UNINTERRUPTABLE POWER SUPPLY

___ Look at anchorage.
___ Ask how unit operated.

EMERGENCY POWER - IMPACT OR LACK OF EMERGENCY POWER

___ If there was a failure of emergency power, what was impact?
___ Where there items or circuits that should have emergency power but did not.
___ What was the impact of lack of emergency power?
___ Was there any loss of emergency power due to the failure of other systems, such as water or cooling.

13. Acknowledgements

13. ACKNOWLEDGEMENTS

Technical editing was done by Anshel J. Schiff. Sections were reviewed by members of the Earthquake Investigations Committee. Sections were prepared by the individuals indicated below.

Section	Authors
1.	Anshel J. Schiff
2.	Anshel J. Schiff
3.	Anshel J. Schiff
4.	Anshel J. Schiff
5.	Anshel J. Schiff
6.	Anshel J. Schiff
7.	LeVal Lund
8.	Holly Cornell
9.	Airports - Anshel J. Schiff
	Harbors - Charles Farrar and Claude Griffin
	Highways - Stuart Werner and Jim Cooper
10.	Alex Tang and Larry Wong
11.	Peter McDonough
	Douglas Honegger
	Douglas Nyman
12.	Tanks - Robert P. Kennedy and Anshel J. Schiff
	Emergency Power - Anshel J. Schiff
13.	Anshel J. Schiff
14.	Anshel J. Schiff

Editorial review of parts of the manuscripts by Otto Steinhardt and Susan K. Welch is gratefully acknowledged.

14. Appendices

    A. Field Guide for Lifeline Earthquake Investigations
        Intensity/Magnitude Scales
             Modified Mercalli Intensity Scale
             Rossi-Forel Intensity Scale
        Departure Check List
        General Rules for Data Collection
        General Site Evaluation Check List
        Anchorage Check List
        Equipment Check List
        Power Systems Check List
            Power Plant
            Power Substation
            Power Equipment
        Water System Check Lists
            Transmission and Distribution
            Pumping Facilities
            Reservoirs
            Ground Water Basins
            Pressure Reduction and Relief Facilities
            Treatment Facilities
            Treatment Plants
        Sewage Systems Check Lists
            Manholes
            Sewers
            Treatment Plants and Pump Pump Station Equipment
            Treatment Plant and Pump Plant Structures
        Transportation System Check Lists
            Airports
            Ports
            Highways
        Communication Systems Check Lists
        Gas and Liquid Fuel System Check Lists
        Tanks Check Lists
        Emergency Power Check Lists

    B. Report Format

    C. Tip on Technical Writing

    D. References to Reconnaissance Reports

Modified Mercalli Intensity Scale

I Note Felt. Marginal and Long-period effects of large earthquakes.
II Felt by persons at rest, on upper floors, or in favorable places.
III Felt indoors. Hanging objects swing. Vibration like passing of light trucks. Duration estimated. May not be recognized as an earthquake.
IV Hanging objects swing. Vibration like passing of heavy trucks; or sensation of a jolt like a ball striking the walls. Standing motor cars rock. Windows, dishes, doors rattle. Glasses clink. Crockery clashes. In the upper range of wooden walls and frames creak.
V Felt outdoors; direction estimated. Sleepers wakened. Liquids disturbed, some spilled. Small unstable objects displaced or upset. Door swing, close and open. Shutters, pictures move. Pendulum clocks stop, start, change rate.
VI Felt by all. Many frightened and run outdoors. Persons walk unsteadily. Windows, dishes, glassware broken, knickknacks, books, etc. fall off shelves. Pictures fall off walls. Furniture moved or overturned. Weak plaster and masonry D cracked. Small bells ring in churches and schools. Trees, bushes visibly shaken or heard to rustle.
VII Difficult to stand. Noticed by drivers of motor cars. Hanging objects quiver. Furniture broken. Damage to masonry D, including cracks. Weak chimneys broken at roof line. Fall of plaster, loose bricks, stones, tiles cornices (also unbraced parapets and architectural ornaments). Some cracks in masonry C. Waves on ponds; water turbid with mud. Small slides and caving in along sand or gravel banks. Large bells ring. Concrete irrigation ditches damaged.
VIII Steering of motor car affected. Damage to masonry C; partial collapse. Some damage to masonry B; none to masonry A. Fall of stucco and some masonry walls. Twisting, fall of chimneys, factory stacks, monuments, towers, elevated tanks. Frame houses moved on foundations if not bolted down; loose panel walls thrown out. Decayed piling broken off. Branches broken from trees. Changes in flow or temperature of springs and wells. Cracks in wet ground and of steep slopes.
IX General Panic. Masonry D destroyed; masonry B seriously damaged. (General damage to foundations.) Frame structures, if not bolted, shifted off foundations. Frames racked. Serious damage to reservoirs. Underground pipes broken. Conspicuous cracks in ground. In alluviated areas sand and mud ejected, earthquakes fountains, sand craters.
X Most masonry and frame structures destroyed with their foundations. Some well-built wooden structures and bridges destroyed. Serious damage t dams, dikes, embankments. Large landslides. Water thrown on banks in canals, rivers, lakes, etc. Sand and mud shift horizontally on beaches and flat land. Rails bent slightly.
XI Rails bent greatly. Underground pipelines completely out of service.
XII Damage nearly total. Large rock masses displaced. Lines of sight and level distorted. Objects thrown into air.

Note: To avoid ambiguity, the quality of masonry, brick, or other material is specified by the following system. (This has no connection with the conventional classes A, B, and C construction.)

Masonry A. Good workmanship, mortar, and design; reinforced, especially laterally, and bound together by using steel, concrete, etc.; designed to resist lateral forces.
Masonry B. Good workmanship and mortar; reinforced, but not designed to resist lateral forces.
Masonry C. Ordinary workmanship and mortar; no extreme weaknesses, like failing to tie in at corners, but neither reinforced nor designed to resist horizontal forces.
Masonry D. Weak materials, such as adobe; poor mortar; low standards of workmanship; weak horizontally.

After Richter, C.F. Elementary Seismology.

MSK Intensity Scale

I Not noticeable   a). The intensity of the vibration is below the limit of sensibility; the tremor is detected and recorded by seismographs only.
II Scarcely noticeable (very slight) a). Vibration is felt only by people at rest in houses, especially on upper floors of buildings.
III Weak, partially observed only. a). The earthquake is felt indoors by a few people, outdoors only in favorable circumstances. The vibration is like that due to the passing of a light truck. Attentive observers notice a slight swinging of hanging objects, somewhat more heavily on upper floors.
IV Widely observed. a). The earthquake is felt indoors by many people, outdoors by a few. Here and there people awake, but no one is frightened. The vibration is like that due to the passing of a heavily loaded truck. Windows, doors and dishes rattle. Floors and walls creak. Furniture begins to shake. Hanging objects swing slightly. Liquids in open vessels are slightly disturbed. In standing motor cars the shock is noticeable.
V Awakening a). The earthquake is felt indoors by all, outdoors by many. Many sleeping people awake. A few run outdoors. Animals become uneasy. Buildings tremble throughout. Hanging objects swing considerably. Pictures knock against walls or swing out of place. Occasionally pendulum clocks stop. A few unstable objects may be overturned or shifted. Open doors and windows are thrust open and slam back again. Liquids spill in small amounts from well- filled open containers. The sensation of vibration is like that due to a heavy object falling inside the building. b). Slight damage of Grade 1 in buildings of Type A is possible. c). Sometimes changes in flow of springs.

VI Frightening   a). Felt by most people indoors and outdoors. Many people frightened and run outdoors. A few persons lose their balance. Domestic animals run out of their stalls. In a few instances, dishes and glassware may break, books fall down. Heavy furniture may possibly move and small steel bells may ring. b). Damage of Grade 1 is sustained in single buildings of Type B, and in many of Type A. Damage in a few buildings of Type A is of Grade 2. c). In a few cases crakes up to widths of 1 cm possible in wet ground; in mountains occasional landslides; change on flow of springs and in the level of well-water is observed.

VII Damage to buildings. a). Most people are frightened and run outdoors. Many find it difficult to stand. The vibration is noticed by persons driving motor cars. Large bells ring. b). In many buildings of Type C damage of Grade 1 is caused; in many buildings of Type B damage is of Grade 2. Many buildings of Type A suffer damage of Grade 3, a few of Grade 4. In single instances landslides of roadway on steep slope, cracks in roads, seams of pipelines damaged, cracks in stone walls. c). Waves are formed on water, and water is made turbid by mud stirred up. Water levels in wells change , and the flow of springs change. In a few cases dry springs have their flow restored and existing springs stop flowing. In isolated instances parts of sandy or gravelly banks slip off.

VIII Destruction of buildings. a). Fright and panic; also persons driving motor cars are disturbed. Here and there branches of trees break off. Even heavy furniture moves and partly overturns. Hanging lamps are in part damaged. b). Many buildings of Type C suffer damage of Grade 2, a few of Grade 3. Many buildings of Type B suffer damage of Grade 3, and many buildings of Type A suffer damage of Grade 4. Occasional breakage of pipe seams. Memorials and monuments more and twist. Tombstones overturn. Stone walls collapse. c). Small landslide in hollows and on banked roads on steep slopes; cracks in ground up to widths of several centimeters. Water in lakes becomes turbid. New reservoirs come into existence. Dry wells refill and existing wells become dry. In many cases changes in flow and level of water.

IX General Damage to buildings. a). General panic; considerable damage to furniture. Animals run to and fro in confusion and cry. b). Many buildings of Type C suffer damage of Grade 3, a few of Grade 4. Many buildings of Type B show damage of Grade 4, a few of Grade 5. Many buildings of Type A suffer damage of Grade 5. Monuments and columns fall. Considerable damage to reservoirs; underground pipes partly broken. In individual cases railway lines are bent and roadways damaged. c). On flat land overflow of water, sand and mud is often observed. Ground cracks to widths of up to 10 cm. on slopes and river banks more than 10 cm; furthermore a large number of slight cracks in ground; falls of rock, many landslides and earth flows; large waves on water. Dry wells renew their flow and existing wells dry up.

X   General destruction of buildings. b). many buildings of Type C suffer damage of Grade 4, a few of Grade 5; critical damage to dams and dikes, and severe damage to bridges. Railway lines are bent slightly. Underground pipes are broken or bent. Road paving and asphalt show waves. c). In ground, cracks up to widths of several tens of centimeters, sometimes up to a meter. Broad fissures occur parallel to water courses. Loose ground slides from steep slopes. From river banks and steep coasts considerable landslides are possible. In coastal areas displacement of sand and mud; change of water level in wells; water from canals, lakes rivers, etc. thrown on land. New lakes occur.

XI Catastrophe b). Severe damage even to well-built buildings, bridges, water dams and railway lines; highways become useless; underground pipes destroyed. c). Ground considerably distorted by broad cracks and fissures, as well as by movement in horizontal and vertical directions; numerous landslips and falls of rocks. The intensity of the earthquake requires to be investigated specially.

XII Landscape changes. b). Practically all structures and below ground are greatly damaged or destroyed. c). The surface of the ground is radically changed, considerable ground cracks with extensive vertical and horizontal movements are observed. Fall of rock and slumping of river banks over wide areas; lakes are dammed; waterfalls appear, and rivers are deflected. The intensity of the earthquake requires to be investigated specially.

Types of Structures
Structure A   Building in field-stone, rural structure, adobe houses, clay houses.
Structure B   Ordinary brick buildings, buildings of the large block and prefabricated type, half timbered structures, buildings in natural hewn stone.
Structure C   Reinforced buildings, well-built wooden structures.
Definition of quantity: Single few - about 5%; Many - about 50%; Most - about 75%

Classification of damage to buildings
Grade 1   Slight damage: Fine cracks in plaster; fall of small pieces of plaster.
Grade 2   Moderate damage: Small crack in walls; fall of fairly large pieces of plaster; pantiles slip off; cracks in chimneys; parts of chimneys fall down.
Grade 3   Heavy damage: Large and deep cracks in walls; fall of chimneys.
Grade 4   Destruction; Gaps in walls; parts of buildings may collapse; separate parts of the building lose their cohesion; inner walls and fill-in walls of the frame collapse.
Grade 5   Total damage: Total collapse of buildings.

Arrangement of the scale   Introductory letters are used to paragraphs throughout the scale as follows: a). Persons and surroundings., b). Structures of all kinds., c). Nature
(From World Data Center A for Solid Earth Geophysics, Report SE-20 1979.)

Rossi-Forel Scale

I Microseismic shock. Recorded by a single seismograph or by seismographs of the same model but not by several seismographs of different kinds: the shock felt by an experienced observer.
II Extremely feeble shock. Recorded by several seismographs of different kinds; felt by a small number of persons at rest.
III Very feeble shock. Felt by several persons at rest; strong enough for the direction or duration to be appreciated.
IV Feeble shock. Felt by persons in motion; disturbance of movable objects, doors, windows, cracking of ceilings.
V Shock of moderate intensity. Felt generally by everyone; disturbance of furniture, beds, etc., ringing of some bells.
VI Fairly strong shock. General awakening of those asleep; general ringing of bells,; oscillation of chandeliers; stopping of clocks; visible agitation of trees and shrubs; some startled persons leaving their dwellings.
VII Strong shock. Overthrow of movable objects; fall of plaster; ringing of church bells; general panic, without damage to buildings.
VIII Very strong shock. Fall of chimneys; cracks in the walls of buildings.
IX Extremely strong shock. Partial or total destruction of some buildings.
X Shock of extreme intensity. Great disaster; ruins; disturbance of the strata, fissures in the ground, rock falls from mountains.

(After Richter, 1956)

Departure Check List

__ Passport, Visa, immunization (vaccination) certifications, and copies
__ Business cards
__ Cash, credit cards, local currency for destination
__ Personal medication, small first aide kit
__ Spare and sun glasses and glasses elastic safety strap
__ Camera, film, spare batteries, and spare camera
__ List of contacts and information on designation
__ US and destination contact points and phone numbers
__ Water bottles, water purification pills, water filter
__ Diarrhea medication
__ Hard hat   __ Gloves   __ Dust mask
__ High energy snack foods
__ Personal clothing
__ Tape recorder, tape and batteries
__ Note pads, clipboard
__ Maps
__ Tape measure, scale, micrometer
__ Magnetic compass
__ Magnifying glass
__ Flashlight with spare batteries
__ Foreign language dictionary
__ Ruler and north arrow for pictures
__ Field Guide
__ Small Shoulder bag, large shoulder bag, suite case

General Rules for Data Collection

- At each site or organization get name, address and phone number of a manager who is familiar with damage repair so that follow-up questions or additional data can be requested.
- When equipment or facilities are damaged, in addition to documenting the damage, attempt to find out 1) failure modes, 2) factors that may have contributed to the failures, 3) implication of the damage on the operation of the facility, 4) implications of the damage on the operation of the system, and 5) resources required to restore service including man days, support equipment, spare parts and total time for restoration. Attempt to gather sufficient details on the equipment and situation so that recommendations to improved the seismic response of the equipment or post-earthquake operations can be made.
- If time is available after investigating damage and the site has experienced, severe ground motions that can be quantified, identify equipment and facilities (similar data that is required when equipment is damaged- see Equipment Check List) that have performed well that are known to have been damaged in past earthquakes.
- Note "tricks" that may be useful to for gathering data in future investigations.
- Pay particular attention to the performance (both good and bad) of new equipment types or designs and new installation practices.

General Site Evaluation Check List

__ What is the orientation of the site relative to magnetic north?
__ What is the topography around the site?
__ What parts of the site are on cut or fill?
__ Can an estimate of the depth of the cut or fill be made by looking at the boundary of the site?
--- Is the site on alluvium? Inquire as to its depth.
--- What are soil conditions at the site: rock, very firm soil, firm soil, soft soil?
--- Type of soil--general non-technical soil types might be
    Soft soil: Loose sand; Unconsolidated silt; Loam; Mud; Dump fill
    Firm soil: Gravel; Consolidated sand; Consolidated silt
    Rock
__ What is drainage of site, above or below grade?
__ What is the depth of the water table?
__ Is there evidence of soil deformations?
__ Does the site manager know if there was any special foundation preparations at the site?
__ Did you get an estimate of the MM Intensity for the site.
__ Did you get a card for the plant manager with his address?
__ Did you evaluate the over all quality of construction and the use of good seismic practices?
__ Did you estimate the percentage of failures for each type of equipment?
__ What was the extent of disruption at this location?
__ What was the duration of the disruption at this location?
__ What was the time (man hours and total) to restore service and to complete repairs? Was special equipment or spare parts needed?
__ What was the impact of dysfunction at this site on system?
__ What was the impact of dysfunction on other lifelines?
__ What was the impact of dysfunction of the lifeline on the emergency response and the community?

Anchorage Check List

__ Are the anchor bolts cast in place or expansion anchors?
__ Can you identify the type or manufacturer of the anchor?
__ What is their length of embedment?
__ What is their diameter?
__ How did they fail?
__ Did they pull out of concrete?
__ Is the concrete cracked? Is it in a tension zone?
__ Did fracture cones develop in the concrete?
__ Did the bolts stretch or pull out slightly so that they are not tight?
__ Did the bolts break?
__ Is there any indication that they were installed incorrectly?
__ What were the standards, if they existed, when the equipment was installed?
__ How many bolts were there and how were they laid out?
__ Did the bolt pass through a structural member in the equipment framing?
__ Are there signs of distress in the equipment in the region around the anchor bolt: cracked or chipped paint, deformation of metal?
__ Does the equipment introduce a prying action to the bolt?
__ Is the bolt hole appropriate to the bolt diameter?
__ Does the load path from the equipment frame to the bolt or weld introduce flexibility in the anchorage system?
__ What are the sources of loading on the anchorage: equipment weight, height of center of gravity, dimensions of the base of the equipment?
__ Were there loads applied through interconnections to adjacent equipment?
__ Has the base of the equipment moved on its footing?
__ Is there a gap around the footing or equipment pedestal indicating differential movement?

Equipment Check List

__ Was there an anchorage failures? If yes, review the Anchorage Check List.
__ Is there a potential for interaction due to the relative motion between adjacent equipment items, such as power cable, control cable, water, oil, fuel line connections? Was connected equipment anchored?
__ Have you followed the load path from the equipment into the supporting structure or soil?
__ Are there signs of any fluid leaks?
__ Have you looked at the interface between the equipment and its support structure? Check for cracked paint, displacements at interface, lose bolts, working of joints as indicated by scratches or burnishing of paint or galvanization.
__ Have you looked at the support structure? Check connections, particularly slotted connections for working of connections.
__ If there is equipment damage get a complete description of the equipment to include the manufacturer, model number, serial number, age, name plate values such as rating and capacity.
__ Get damage statistics for equipment that is damaged, that is, get the total number of similar equipment items and the percentage that has been damaged.
__ For similar equipment that is at a site in which some is damaged and some is not, can an explanation of the different performance be found?
__ Check equipment with rotating or moving parts that might bind due to deformations.

Power Systems - Power Plant Check List

- __ Review Site Check List
- __ Inquire about damaged equipment.
- __ Look for interaction problems between the boiler support structure and the boiler.
- __ Look for interaction problems between the turbine pedestal and the powerhouse operating floor.
- __ Inquire if the unit when off line. If so, determine why.
- __ Are there any indications of turbine bearing damage?
- __ Does there appear to be steam coming from the stack indicating boiler tube damage?
- __ Generally, how is equipment anchored?
- __ Check for damage to coal handling equipment, in particular, conveyor anchorage points.
- __ Check station batteries.
- __ Were sudden pressure relays in transformers activated?
- __ Did any protective relays change state? Which ones and what was impact?
- __ Were any relays reset after the earthquake to resume operations?
- __ Was there a loss of power on any lines into or out of the station?
- __ If there were any disruptions, what were the causes and what was the duration?
- __ If there was a suspended ceiling, did any of the panels fall?
- __ Did any thing fall from desks, tables or shelves in the substation?
- __ Were there any disruption in communications? If so, what types of communications are used and which were effected? In particular check on radio, microwave, leased telephone lines and regular telephone lines.
- __ Have personnel that were on the site at the time of the earthquake describe the earthquake and their actions after the earthquake.
- __ Are the personnel aware of any other effects that the earthquake had on the power system?
- __ Check Substations Check List in reviewing the switchyard.

Power Systems - Substation Check List

__ Review Site Check List
__ Check for damaged equipment.
__ For vulnerable equipment (circuit breakers, lightning arresters, transformers, current and potential transformers, capacitor racks, and line traps) review Power Equipment Check List.
__ Check station batteries.
__ Were sudden pressure relays in transformers activated?
__ Did any protective relays change state? Which ones?
__ Were any relays reset after the earthquake to resume operations?
__ Was there a loss of power on any lines into or out of the station?
__ If there was any disruptions, what was the cause and what was the duration?
__ If there was a suspended ceiling, did any of the panels fall?
__ Did any thing fall from desks, tables or shelves in the substation?
__ Were there any disruption in communications? If so, what types of communications are used and which were effected?
__ Have personnel that were on the site at the time of the earthquake describe the earthquake and their actions after the earthquake.
__ Are the personnel aware of any other effects that the earthquake had on the power system?

Power Equipment Check List

__ What slack in power connections to adjacent equipment equipment?
__ Is there a potential for interaction due to the relative motion between adjacent equipment items, between an equipment item and bus support, or due to dynamics of flexible bus?
__ Have you followed the load path from the equipment into the supporting soil?
__ Have you looked at the base of bushings for signs of oil leaks, displacements of the gaskets, or displacement of bushings?
__ Have you looked at the interface between the equipment and its support structure? Check for cracked paint, displacements at interface, lose bolts, working of joints as indicated by scratches or burnishing of paint of galvanization.
__ Have you looked at the support structure? Check connections, particularly slotted connections for working of connections.
__ Review the Anchorage Check List.

Water Systems Check List - Transmission and Distribution Systems

__ Get statistics on the transmission and distribution systems - for each of the categories below, get total length of each type of pipe, percentage in areas with significant damage and soil deformations and the number of failures per unit length. Get details on the types of failures and associated statistics. This would include type of "break" such as guillotine break, crack, failure at joint (and type of joint) did corrosion play an important role in the break, was the break near a "T" or bend in pipe or a connection at a small pipe connection? Make up a list to leave with the organization indicating the type of deaggregation that you desire and request that it be sent after the data is collected. Performance of plastic pipe of of particular interest.
__ Pipe size
__ Pipe wall thickness--this may be difficult to see directly and may need to be obtained from operating personnel
__ Pipe material--common materials are vitrified clay, concrete, asbestos cement, plastic (pcv or pe), cast iron, ductile iron, concrete cylinder, and steel.
__ Type of joint--joint types in common use are bell and spigot with cement mortar, bell and spigot with rubber ring, tongue and groove with cement mortar, tongue and groove with rubber ring, push-on and many proprietary types
__ Age of pipe or date of installation
__ Depth of bury
__ Type of soil--general non-technical soil types might be
    Soft soil: Loose sand; Unconsolidated silt; Loam; Mud; Dump fill
Firm soil: Gravel; Consolidated sand; Consolidated silt
Rock
__ Type of bedding material--if special bedding was provided (this information will probably have to be obtained from the operating or maintenance personnel or from the specifications for the original construction
_ Type of backfill - sometimes special backfill material is used up to the spring line or a few inches over the top of the pipe and this information will also have to come from operating or maintenance personnel or the original plans and specifications
--- Ground water elevation
--- Drainage -- surface or underground
--- Description of break or damage
--- Preliminary evaluation of cause of damage
--- Surface indications of ground movement including extent of Faults; Uplift; Subsidence; Slides; Lateral spreading; Soil cracks--size and extent; Other information of interest in analyzing the failure.

Water System Check List - Pumping Facilities

- ___ Description
- ___ Foundation
- ___ Power supply
- ___ Piping
- ___ Well equipment
- ___ Potential sources of pollution
- ___ Operational status
- ___ Building or enclosure status
- ___ Emergency power supply
- ___ Emergency battery rack
- ___ Controls
- ___ Fuel storage

Water System Check List - Reservoirs

- ___ Description
- ___ Determine water level at EQ time
- ___ Upstream and downstream surface
- ___ Spillway, inlet and outlet
- ___ Size and capacity
- ___ Operational status
- ___ Right and left abutments
- ___ Hydroelectric facilities
- ___ Note if Federal or State dams authorities making detailed investigation

Water System Check List - Ground Water Basins

- ___ Description
- ___ Levees
- ___ Operational status
- ___ Direction of flow
- ___ Flow structures
- ___ Potential sources of pollution

Water System Check List - Pressure Reducing and Relief Facilities

- ___ Description
- ___ Vault condition
- ___ See check list under Pumping Facilities
- ___ Mechanical, electrical and hydraulic controls
- ___ Operational status

Water System Check List - Treatment Facilities

- ___ Description
- ___ Foundation
- ___ See check list under Pumping Facilities
- ___ See check list under Sewerage Systems (Chapter 8.0) Wastewater
- ___ Operational status

Water System Check List - Treatment Plants

Services
- ___ Description
- ___ Damage
- ___ Bursting
- ___ Operational status
- ___ Plugging

Control Center
- ___ Description
- ___ Building status
- ___ Operational status
- ___ Communication lines

Service Yards
- ___ Description
- ___ Building status
- ___ Operational status

Sewage Systems Check List - Manholes

__ Diameter of ring and cover
__ Size--usually diameter below the cone
__ Depth--edge of rim to flow line
__ Type--best described by sketch
__ Material--brick, concrete block, precast concrete, cast in place concrete
__ Inlet pipe size
__ Outlet pipe size
__ Laterals or drops--size and location
__ Nature of flow--free flowing or stagnant
__ Depth of flow--above flow line
__ Plumbness - estimated out of plumb in inches per foot in direction of main sewer and at right angles
__ Structural damage, cracks or breaks

Sewage Systems Check List - Sewers

__ Pipe size
__ Pipe wall thickness--this may be difficult to see directly and may need to be obtained from operating personnel
__ Pipe material--common materials are vitrified clay, concrete, asbestos cement, plastic, cast iron, ductile iron, concrete cylinder, and steel
__ Type of joint--joint types in common use are bell and spigot with cement mortar, bell and spigot with rubber ring, tongue and groove with cement mortar, tongue and groove with rubber ring, push-on and many proprietary types
__ Age of pipe or date of installation
__ Depth of bury
__ Type of soil--general non-technical soil types might be
__ Soft soil: Loose sand; Unconsolidated silt; Loam; Mud; Dump fill
   Firm soil: Gravel; Consolidated sand; Consolidated silt
   Rock
__ Type of bedding material--if special bedding was provided (this information will probably have to be obtained from the operating or maintenance personnel or from the specifications for the original construction
__ Type of backfill - sometimes special backfill material is used up to the spring line or a few inches over the top of the pipe and this information will also have to come from operating or maintenance personnel or the original plans and specifications
--- Ground water elevation
--- Drainage -- surface or underground
--- Description of break or damage
--- Preliminary evaluation of cause of damage
--- Surface indications of ground movement including extent of Faults; Uplift; Subsidence; Slides; Lateral spreading; Soil cracks--size and extent; Other information of interest in analyzing the failure.

Sewage Systems Check List - Treatment Plants and Pump Stations Equipment

__ Interview operators for failed equipment
__ Determine or verify for failed equipment
   Make; Model; Size ; Capacity
__ Verify reasons for shutdown
__ Look for secondary causes of failure
   Lack of power; Piping damage; Control system failure; Fuel line damage; Lubricating system failure; Other causes
__ Review anchorage (see Anchorage check list)
__ Attempt to determine cause of failure
__ Seek assistance from manufacturer, maintenance personnel, designers, consultants.

Sewage Systems Check List - Treatment Plant or Pump Station Structures

__ Interview operator and maintenance personnel
__ Examine joints
__ Survey for cracks and broken supports
__ Check alignment of walls, and channels
__ Look for differential movement between structures
__ Investigate pipe and conduit connections to structures
__ Evaluate pipe hangar performance
__ Sketch or describe movements causing damage
__ Make a preliminary determination of the cause of the failures

Airport Check List

Basic Data:
- __ No. and length of runways
- __ Passenger and freight traffic
- __ Map

Airfield-Side
Control Tower:
- __ power
- __ equipment
- __ radio communication
- __ windows
- __ ceiling
- __ telephone communic.

Air Traffic Control:
- __ radars
- __ approach lights
- __ Runway and taxiway damage
- __ beacons
- __ ground control radios

Fuel Supply:
- __ Tanks
- __ Piping
- __ Power for tank farm

Status of aircraft service facilities:
- __ Freight facilities
- __ Freight storage facilities
- __ Aircraft food service

Communications:
- __ Radio saturation
- __ Repeater operability, emergency power
- __ Telephone operation, phone saturation, on-field and off-field communications
- __ Passenger Baggage Facilities
- __ Emergency response facilities: power, security, fire, medical

Land-Side

Terminal:
- __ Structure
- __ Jetways
- __ Non-structural damage: ceilings, fire suppression
- __ Power

Communications:
- __ Public address system
- __ Airport service phones (courtesy, security, maintenance)
- __ On- and off- field telephone communications
- __ Contact with emergency response
- __ Operation of 911
- __ Emergency power for communications
- __ Airport based transportation systems
- __ Airport-community transportation links
- __ Emergency power

General
- __ Water
- __ Sewage systems

Impacts on Operations
- __ Flight flag-offs and diversions
- __ Passengers
- __ Personnel problems
- __ Evacuations

PORT CHECK LIST

General Information
- Names, addresses, phone numbers, and titles of contact persons.
- Organizational management (private, municipal, district, etc.).
- Type and amount of cargo handled by the harbor, annual revenues.
- If ferries use the facility, the number of passengers per day before and after the earthquake.
- Estimate (total dollar amount) of damage and lost revenues.
- Measured acceleration levels.

When investigating a harbor facility, we are typically interested in obtaining a brief description of the portions of the facility listed below, including the conditions of these facilities (any existing damage or damage caused by past seismic events) prior to the earthquake. Next, we would like to document the damage sustained by these portions of the harbor, the impact of that damage on the users, the methods (both permanent and interim) used to restore service, and obstacles (lack of equipment, personnel or materials, lack of access) to the restoration of services.

Navigation Channel
- Information on changes in Channels
- Breakwaters, Jetties
- Seawalls, Quay Walls, Bulkheads, Revetments
- Dikes
- Other retaining structures

Piers and Wharfs
- Support structures (piles, caissons, fill, etc.)
- Deck
- Pier or wharf/shoreline interface
- Utility lines buried in the deck
- Seawalls and other retaining structures

Cargo Handling Equipment
- Container cranes
- Conveyors
- Chiksans/pipelines
- Specialty equipment
- Other cranes
- Stacker and reclaimer equipment
- Ramps

Cargo Storage Areas
- Warehouses
- Silos
- Specialty facilities
- Open lots (asphalt, concrete, unpaved)
- Tank farms
- Container control tower

Inland Transportation Systems
- Roads
- Rail lines
- Pipelines
- Roadway bridges
- Railroad bridges

Interfaces with Other Lifelines
- Power
- Water
- Sewage
- Natural gas
- Communication systems

Other Buildings
- Administration and office buildings
- Fire stations
- Maintenance/repair buildings

Emergency Plans
- Describe the plan (if any).
- How did the plan perform?
- Based on the experiences gained from this earthquake, will changes or improvements be made to the plan?

HIGHWAY CHECK LIST

\multicolumn{3}{c}{CHECKLIST FOR POST-EARTHQUAKE INSPECTION OF HIGHWAY ROAD/BRIDGE SYSTEMS}		
ELEMENT	CHECKLIST ITEM	POSSIBLE CAUSE(S)
Roadway	Localized cracking, settlement	Settlement, heaving, and/or liquefaction of backfill or underlying soil materials
	Heaving of adjacent roadway	
	Sandboils in adjacent soil materials	Soil liquefaction
	Widespread and deep extensional cracking and/or slumping	Surface fault rupture
	Sloughing/sliding/rockfalls of adjacent hillsides or slopes	Sliding due to liquefaction and/or to strong ground shaking
Bridge Abutments	Rigid body settlement or lateral movement of abutments	Damage to abutment foundation
	Sandboils in adjacent soil materials	Soil liquefaction
	Rigid body tilting of end wall	Excessive lateral pressure in backfill due to pore pressure buildup or strong ground shaking
	Cracking of abutment walls	
	Separation of soil backfill from abutment walls	Damage to abutment foundation
Bridge Column/Pier Supports	Tilting, settlement, horizontal movement of supporting foundation	Damage to pier foundation
		Movement of subsurface soil materials due to pore pressure buildup or strong ground shaking
	Lateral cracking at connection to bent beam, permanent deformation or rotation at connection to deck	Insufficient strength and/or ductility of beam-pier connection
	Diagonal cracking at pier column connection (at multi-column bent)	Insufficient shear resistance at connection
	Diagonal cracking, spalling, along length of column	Insufficient shear reinforcement
	Transverse cracking of column near top or bottom supports	Insufficient longitudinal reinforcement
Bridge Deck	Sliding, scratchmarks, rubbing, spalling, permanent movement at expansion joint	Longitudinal forces that exceed frictional resistance at joint
	Vertical transverse cracking	Insufficient longitudinal reinforcement
	Diagonal cracking	Insufficient shear reinforcement

Communications Check List - Power Room

__ Battery rack - are batteries confined to rack, is there filler between cell and restraint. Are the racks braced?
__ Cell jar integrity - look for cracks in the cell wall and top near connectors and for internal damage
__ Intercell connectors
__ Anchor condition
__ Frame condition

Communications Check List - Rectifiers and Distribution Panels
__ Cabinet condition - look at anchorage over stress, interaction due to cabinet motion (particularly for tall cabinets), lose circuit backs.
__ Anchor condition - for damaged equipment see equipment and anchorage check lists.
__ Power cable connections - look for pinching and damage due to lack of slack.
__ Output

Communications Check List - Bus Bar
__ Cable connections   __ Splice joints
__ Buckling or short   __ Mechanical support

Communications Check List - Engine Generators
__ Engine skid mounts   __ Intake air and exhaust pipes
__ Generator housing   __ Start battery rack
__ Start air tanks & compressors __ Oil or water leaks
__ Power output cables   __ Outside radiator
__ Fuel and water lines - check day and storage tanks
__ Check for malfunctions in control panels due to inappropriate relay actions.

Communications Check List - Distribution Frame

Frame Structure
__ Mechanical integrity
__ Anchor condition
__ Overhead supports - deformation of overhead frames and failure or framing anchorage.

Terminal Blocks
__ Wiring condition   __ Block mounting
__ Protector capsules

Communications Check List - Switch Room

Switch Equipment
__ Functional capability   __ Cabinet integrity
__ Anchor condition   __ Circuit board walkout
__ Alarm condition   __ Back cabling

Overhead Ironwork
- __ Falling hardware   __ Cable condition
- __ Ceiling anchor condition
- __ Mechanical integrity - check for deformation and anchorage failure.

Switch Room Operating Area
- __ Desk top equipment condition   __ Data cabinet condition
- __ Spare parts storage

Communications Check List - Transmission Room

Transmission Equipment
- __ Functional capability   __ Mounting rack integrity
- __ Anchor condition   __ Circuit board walkout
- __ Alarm condition   __ Overhead bracing condition
- __ Back cabling

Overhead Ironwork
- __ Falling hardware   __ Cable condition
- __ Ceiling anchor condition
- __ Mechanical integrity - check for deformation and anchorage failure.

Transmission room control area
- __ Desktop equipment   __ Spare parts storage

Communications Check List - Building Facilities

Structure
- __ Wall Condition   __ Ceiling condition
- __ Floor condition   __ Staircases
- __ Joints   __ Outside walls

Elevators
- __ Operating status   __ Landing condition

Heating, Ventilation and Air Conditioning
- __ Operating status   __ Mechanical integrity
- __ Pipe hangers   __ Heat exchanger
- __ Coolant leaks   __ Duct condition
- __ Power panel   __ Check anchorage
- __ Check all utility connections (water, power, control, fuel, air, etc.) for damage due to motion of equipment.

Communications Check List - Data Room

Data Equipment
- __ Equipment operation   __ Cabinet integrity
- __ Cable condition   __ Disk drive condition

Raised Floor
- __ Pedestal condition   __ Floor panel condition

Power Distribution
- __ Distribution panel condition __ UPS battery condition

Air Conditioning
__ Cabinet integrity   __ Water pipes
__ Cabinet bracing and anchors   __ Operation

Communications Check List - Cable Vault

Cable Supports
__ Cable condition   __ Wall anchors
__ Outer wall opening condition

Pressurization Equipment
__ Operation   __ Cabinet condition
__ Anchor condition
__ Tube connections

GAS AND LIQUID FUEL LIFELINE ISSUES CHECK LISTS

Vulnerability
__ Impact of lifeline failure on public
__ Repair and restoration costs
__ Restoration time
__ Economic impact to customers

Regulatory Policy
__ Appropriate level of protection
__ Regulatory actions
__ Funding
__ Standards

Earthquake Awareness
__ Lesson learned
__ Design/operations guidance

Design
__ Pipeline performance at high strain
__ Soil-pipeline interaction characteristics

PIPELINE FAILURES CHECK LIST

If possible, all piping information should be recorded by individual pipe size and material.

General Conditions
__ Miles of pipe in impacted areas
__ Operating pressure
__ Maximum allowable pressure
__ Size (diameter)
__ Wall thickness
__ Material
__ Pipe grade or specification

Type of Failure
__ Shear   __ Compression
__ Crack   __ Joint pull-out

Location of Failure
__ Pipe   __ Weld or fusion
__ Coupling   __ At bends, tees, valves, etc.

Failure Consequences
__ Fire   __ Explosion
__ System shutdown or underpressure

Probable Cause of Failure
__ Liquefaction   __ Faulting
__ Ground shaking   __ Landslide

Condition of Pipeline
__ Corrosion					__ Type of weld or connection

Soil Conditions
__ Soil Type					__ Depth to pipe
__ Ground Water					__ Special bedding

FACILITY

Type
__ Pump station					__ compressor station
__ Metering station				__ Pressure regulator station
__ Product storage

Location
__ Above ground					__ Vault
__ Building (material)

Interruption of Operation
__ External causes				__ Component failure

Facility Performance
__ Equipment					__ Piping
__ Structure					__ Instrumentation and control
__ Tankage

Failure Consequences
__ Fire						__ Explosion
__ System overpressure				__ System shutdown or
						    underpressure

CUSTOMER RELATED EQUIPMENT CHECK LIST

Type of Equipment
__ Metering					__ Domestic appliances
__ Commercial equipment				__ Industrial Equipment

Type of Damage
__ Piping					__ Inadequate equipment
						    anchorage

Failure Consequences
__ Fire						__ Explosion
__ System overpressure				__ System shutdown or
						    underpressure

Emergency Devices
__ Manual valves				__ Automatic shut-off valves
__ Did customer know how to shut down system?

OPERATIONS AND MAINTENANCE CHECK LIST

Service Interruptions
__ Number of customers   __ Duration of interruption
__ Reduction of throughput

Maintenance and Repair
__ Number of repair orders   __ Numbers and types of emergencies
__ Total cost or man-hours   __ Availability of spare parts
__ Long term increase in leaks

Earthquake Preparedness Assessment
__ Employee training   __ Emergency response plan
__ Internal communications   __ External liaison
__ Maps and records

Tank Check List - Vertical Flat Bottom Tanks

General
__ Facility identification of tank
__ Are detailed tank drawings available?
__ Type of liquid in tank - density and viscosity

Foundation
__ Describe general site conditions - settlement, slope instability, liquefaction (review Site Condition Check List)

Geometry
__ Tank Diameter __ Height to top of wall
__ Height of overflow outlet __ Fluid height at EQ time
__ Type of top, Flat, Domed, Floating
__ Material - carbon steel, aluminum, stainless steel
__ Type of wall construction - welded, riveted
__ Wall thickness near base __ Wall thickness above base
__ Existence of ring stiffeners and location (Clues to internal stiffeners?)
__ Type of support - ground, ring skirt, legs
__ Thickness of scetch plate
__ Type of weld at scetch plate - single or double fillet
__ Type of foundation - soil, asphalt, concrete
__ Condition of foundation - flat, domed, cracks, subsidence
__ Type of anchorage - bolts, straps, retaining ring
__ Spacing of anchorage (on centers) or number of anchors
__ Detailed description of anchor
   Strap width and thickness
   Bolt diameter, length above grade, total length, chair detail
__ Piping connections - diameter, distance to anchor (flexibility), valves
__ Type of drain - above grade, below grade
__ Vents, type and diameter
__ Height of overflow outlet
__ Height of liquid in tank at time of earthquake
__ Type of liquid in tank - density and viscosity
__ Review Site Conditions Check List

Signs of Distress
__ Signs of lateral motion - size, direction
__ Failure of drain pipe below tank bottom due to tank motion
__ Failure at piping connection due to lateral motion of tank
__ Signs of base lift - stretched, bulled, or broken bolts distressed straps, disturbed ground
__ Elephant foot - height (top to bottom), position of bottom to base, increase in radius, extent around tank, distress of bolts at opposite side, orientation of buck relative to magnetic North, condition of weld
__ Base plate weld failure - length, indications of corrosion, quality of weld
__ Signs of wall distress - chipped paint, leaks, buckles, location
__ Failure of pipe connections
__ Buckling at top of tank - sloshing, implosion
__ Buckling of roof
__ Buckling at stiffener near to of tank
__ Signs of leaks or spills

Horizontal Tanks

__ Tank diameter and length
__ Material - carbon steel, aluminum, stainless steel
__ Thickness
__ Shape of tank ends
__ Supported on saddles or legs - number of supports
__ Details of supports - type of material, dimensions
__ Provisions for longitudinal stiffness of supports
__ Details of attachment of tank to supports
__ Indications of tank stiffeners near supports
__ Anchorage of supports to foundation - See Anchorage Check List
__ Details of provision for longitudinal expansion of tank
__ Type of foundation
__ Pipe connections - diameter, flexibility
__ Fullness of tank at time of earthquake, type of material
__ Review Site Check List

Signs of Distress

__ Motion of tank in saddle __ Distress in supports
__ Distress at support anchorage __ Action at slip connections
__ Signs of distress - scraped paint, chipped paint, deformation of members

EMERGENCY POWER - BACK-UP BATTERIES

__ Is battery rack anchored, braced, and can it withstand lateral loads?
__ Are batteries restrained with spacers between cells, end and side restraints?
__ Look for damage to cell case, post, plates, and bus bars.
__ Look at anchorage and condition of chargers and inverters - check internal transformers.

EMERGENCY POWER - ENGINE-GENERATORS

__ Anchorage of engine-generator (E/G) - snubbers
__ Check slack and stiffness of utility lines to E/G - power, fuel, water, air, and control lines.
__ Check exhaust and intake air ducts.
__ Check day tank and its fuel lines.
__ How is fuel supplied to the day tank - is power needed?
__ Check control cabinet and ask about its function.
__ Check security of starting batteries and connections (for battery start units).
__ Check motor-compressor, isolation, lines, and tank (for air start units).
__ If the unit feeds a unit substation, look at anchorage of dry-type transformer.
__ Ask if unit started and operated as expected.
__ Ask if and how unit is tested - frequency, under load, and how is it test started.

EMERGENCY POWER - UNINTERRUPTABLE POWER SUPPLY

__ Look at anchorage.
__ Ask how unit operated.

EMERGENCY POWER - IMPACT OR LACK OF EMERGENCY POWER

__ If there was a failure of emergency power, what was impact?
__ Where there items or circuits that should have emergency power but did not.
__ What was the impact of lack of emergency power?
__ Was there any loss of emergency power due to the failure of other systems, such as water or cooling.

APPENDIX B REPORT FORMAT

A. Contents for TCLEE Reports (The ordering of the various lifelines is arbitrary. The contents of each section and subsection listed below should follow that given under B. below.)

Introduction
    Executive Summary
    Organization
    Acknowledgements
    Chronology

Table of Contents

Geoscience, Geotechnical, and Strong Motion Data

Water and Sewage Systems (These may be separate sections.)

Transportation
    Bridges and Highway Structures
    Highways
    Railways
    Airports
    Ports

Communication
    Telephone - Public Switched Network
    Cellular Telephones
    Public Safety Communications - 911 Systems
    Radio Communications
    Hospital Radio Network
    Amateur Radio Emergency Response (Ham Operators)
    Public Media - Print, Broadcast, Emergency Response Network

Power System

Gas and Liquid Fuel System

Recommendations for improving earthquake investigations.

Summary and Conclusions

Appendices

B. Contents For Each Lifeline System

This format would be followed for each lifeline under the major sections, ie, airports, ports and highways under transportation.

Introduction
    An Overview
    Limitations and general disclaimers associated with the investigation
    Organization of report on that lifeline

Each Lifeline would be organized along these lines.

    System Description

An overview of the system, its size, service area, types of facilities (for example types and sizes of pipe, operating voltages, system map, etc.)

Description of Damage

A facility by facility listing of the damage, and for each damaged item, indicate possible failure modes and factors contributing to the failures. Pictures should be informative, necessary to understand the failure and have some degree of uniqueness. For example, an elephant foot tank would probably be inappropriate as there are typically several per earthquake. However, a detail of a damaged tank anchor bolt chair would be of interest.

For some lifelines, for example, water system piping, the details of each failure would, in general, not be of interest, but statistics of failures and conditions associated with them would be important.

Immediate Post-Earthquake Response

Describe how the system came down and interactions between lifelines immediately after the earthquake. This section would describe the response of the system in the minutes and hours after the earthquake. In this and the next section, comparisons noting similarities and differences relative to previous earthquake should be made.

Post-Earthquake Recovery

This would describe how the system was restored.

Observations and Recommendations

The observations would indicate damage and those aspects of the impact of the earthquake which were new and observed previously with an emphasis on new observations. Special problems should be noted. Recommendations should be given by a short one line statements followed by a paragraph with amplification as needed. They should be based on observations from the event.

APPENDIX C TIP FOR TECHNICAL REPORT WRITING

THE COLON

The colon (:) is a connective form of punctuation signifying that more information is to follow. A colon communicates a relationship between the information that precedes it and the information that follows: the part of the sentence before the colon acts as an introduction; the part of the sentence following the colon serves as an explanation or example. Use it to make your writing more dynamic.

In technical writing, a colon is often followed by a series.

- There are three significant faults in the area: The San Andreas, the Hayward, and the Calaveras.

A colon can be used to combine two sentences into one when both sentences contribute to the expression of the same idea.

- The climate is not suitable for field work: temperatures often drop below freezing.

Colons usually introduce tabulations. In running text, however, the colon usually is not used after such introductory words.

- Send the information to:
- The following are recommended:

The colon is also used after the salutation in a formal letter (Dear Mr. Smith:), in writing the time (6:45:19), in separating chapter and verse in Bible citations (Luke 3:7), and in expressing ratio (1:100).

When typing, space twice after a colon. No dash should be used.

HYPHEN

The hyphen (-) is generally used to connect compound adjectives (north-trending) and one-thought expressions (right-of-way). Its use is a matter of judgment aided by a few guidelines listed below. The main rule is to use it consistently throughout a report.

- Hyphenate words that together act as an adjective (fine-grained sand; state-of-the-art study; flood-plain deposits; blue-gray clay). However, do not hyphenate these words when they follow a verb and no longer act as a unit (sand is fine grained; study represents the state of the art; deposits are on the flood plain; clay is blue-gray). (Obviously, awkward hyphenation can be remedied by rearranging the sentence.)

- If the first word of a multiple-word adjective is an adverb ending in "ly" or an added adverb changes the relationship, the words are not hyphenated (poorly sorted sand; very fine grained sand).

- Hyphenate numeral modifiers (3-inch crack; 4-mile-wide zone; four-part report).

- Combined forms, prefixes, and suffixes do not ordinarily require hyphens (subbottom; anticline; volcaniclastic). Exceptions are ex-, self-, all-, and quasi-, and are made to avoid tripling of letters (self-propelled; all-inclusive; shell-like).

There is a uniform scheme for hyphenating petrologic terminology. Names within any one class are hyphenated, others are not. The four classes are rock names, mineral names, textural terms, and names describing clastic aggregation (biotite-pyroxene andesite; alb1te-epidote-chlorite schist; porphyritic nepheline syenite).

THAT VERSUS WHICH

In The Elements of Style, William Strunk, Jr. writes, "That is the defining or restrictive, pronoun; which is the nondefining, or nonrestrictive." He gives the following examples:

> The lawn mower that is broken is in the garage.
> (tells which one)

> The lawn mower, which is broken, is in the garage.
> (adds a fact about the only lawn mower in question)

That and which are not interchangeable. It is appropriate to use which when intending a parenthetical remark about something. Which is preceded by a coma. That is not preceded by a comma and starts a phrase that provides crucial information. To determine proper usage, try deleting the phrase. In the second example, "The lawn mower is in the garage" imparts the basic meaning intended; "which is broken" just adds information. The first sentence is about a particular lawn mower and "that is broken" cannot be deleted.

This distinction is often important in technical writing (less so in creative writing). Notice the difference between "the faults, which are active," and "the faults that are active." Improve your technical communication by conducting a "which" hunt.

SEMICOLON

The semicolon is a stop. It is a stronger stop than a comma; it does not have the force of a period. In these cases the semicolon is a connector; it is used instead of the connecting words "but" and "and".

The semicolon serves a coordinating function and adds variety to writing by allowing one to leave out words and still retain meaning. To some writers, the semicolon is a useful tool; to others, a nuisance.

The use of the semicolon can be avoided; nevertheless, it provides emphasis and clarity that would be difficult to achieve in another way, especially when thoughts are closely related, as they are in this sentence. (Nevertheless is a connecting word and could start a separate sentence; however, the semicolon emphasizes the connector, and leaves no doubt that the last thought relates to the beginning of the sentence).

The semicolon is also used before starting an explanation or enumeration; for example, one, two, and three, or; namely, red, yellow, and blue.

For clarity, the semicolon is substituted for commas when listing a series of phrases, especially when the phrases themselves have commas in them.

DASH

The dash (--) separates elements of a sentence--words, phrases, clauses. It is used to set off highly parenthetical matter, to abruptly separate independent clauses, and to indicate special emphasis, interruption, or shift in thought. Dashes denote informality and should not be used indiscriminately for commas.

Dashes are usually used in pairs to set off material completely; commas, semicolons, and colons may be used within the material set off by dashes and do not terminate the influence of the dash. The dash is strong punctuation and its authority is ended only by a period, the close of a parenthesis, or another dash.

- These are shore deposits--gravel, sand, and clay--and marine sediment underlies them.

The dash is typed as two hyphens and is typeset as a single long line. No spaces are used before or after the dash, and commas, colons, or semicolons are not used immediately preceding or following the dash. A dash that falls at the end of a line of type is placed on that line and not at the start of the next.

Avoid using the dash as a replacement for words in phrases such as "4 to 6" and "1906 through 1974" (1906-74 cannot be precisely translated). The dash is seldom needed in a formal report or proposal and should not be used as a substitute for a properly constructed, coordinated sentence.

ACTIVE VS PASSIVE VOICE

The passive voice is a verb form in which the subject of the sentence is being acted upon. In the active voice, the subject of the sentence acts. The verb form most commonly used in technical writing is the passive voice because it emphasizes the thing being acted upon, rather than the actor. Thus, we say: *The faults in the foothills were studied.* rather than *The geologist studied the faults in the foothills.*

The convenience of the passive voice for reporting information has lead to habitual use of this form, even when it is not necessary. The most direct statements use the active voice and your writing will be more interesting if you take advantage of opportunities to use it. Do not say *Thirty gallons per minute are discharged by the pump.* when you can say *The pump discharges 30 gallons per minute.* Change *The scheduled scope of work was not completed by the contractor* to *The contractor did not complete the scheduled scope of work. Water displaces oil* is much more natural than *the displacement of oil by water occurs.*

Do not shift between active and passive voices in the same sentence.

Not: *Microseismic studies cost little and much is learned from them.*

But: *Microseismic studies cost little and provide much data.*

Not: *We visited the site, where many landfalls were seen.*

But: *We visited the site, where we saw many landfalls.*

APPENDIX D. REFERENCED TO RECONNAISSANCE REPORTS

The Great Alaska Earthquake of 1964," Committee on the Alaska Earthquake, National Research Council, National Academy of Sciences, Washington, D.C., 1973

San Fernando, California, Earthquake of February 9, 1971," U.S. Dept. pf Commerce, NOAA, Vol. II, Washington, D.C. 1973.

"Chile Earthquake of March 3, 1985 - Performance of Lifelines," Earthquake Spectra, Vol. 2, No. 2 1986, pp. 429-482.

"Lifelines," Morgan Hill Earthquake of April 24, 1984, Earthquake Spectrum, Vol. 1, No. 3, 1985, pp 615-666.

"Reducing Earthquake Hazards: Lessons Learned from Earthquakes, EERI, Publication 86-02, Nov. 1986, pp. 83-114.

"The Whittier Narrows Earthquake of October 1, 1987 - Response of Lifelines and Effects on Emergency Response" A. J. Schiff , Earthquake Spectra, Vol. 4, No. 2, May 1988, pp. 339-366.

"Lifeline Response to the Tejon Ranch Earthquake," A.J. Schiff, Earthquake Spectra, Vol. 5, No. 4, November 1989, pp. 791-812.

"Loma Prieta Earthquake Reconnaissance Report - Lifelines," A.J. Schiff with others, Earthquake Spectra, Supplement to Vol. 6, May 1990.

Philippine Earthquake Reconnaissance Report - July 16, 1990, to be published.